OPTIMAL RECOVERY OF ANALYTIC FUNCTIONS

OPTIMAL RECOVERY OF ANALYTIC FUNCTIONS

K. YU. OSIPENKO

Nova Science Publishers, Inc.
Huntington, New York

Editorial Production:	Susan Boriotti
Office Manager:	Annette Hellinger
Graphics:	Frank Grucci and Jennifer Lucas
Information Editor:	Tatiana Shohov
Book Production:	Donna Dennis, Patrick Davin, Cathy DeGregory, and Lynette Van Helden
Circulation:	Latoya Clay, Anna Cruz, and Lisa DiGangi

Library of Congress Cataloging-in-Publication Data
Osipenko, K. Yu.
Optimal Recovery of Analytic Functions / K. Yu. Osipenko
p. cm.
Includes index.
ISBN 1-56072-821-3
1. Analytic functions. 2. Approximation theory. I. Title
QA331.O85 2000
515'.73—dc21 00-041146

227 Main Street, Suite 100
Huntington, New York 11743
Tele. 631-424-6682 Fax 631-424-4666
e-mail: Novascience@earthlink.net
e-mail: Novascil@aol.com
Web Site: http://www.nexusworld.com/nova

Printed in the United States of America

To Xenia and Kirill

Contents

Preface

The present book is devoted to the problems of optimal recovery on classes of analytic functions. We consider the problems of optimal interpolation, differentiation, and quadrature formulas. Many problems of Approximation Theory may be considered as optimal recovery problems. Those problems are Kolmogorov's inequality and n-widths which we study for the Hardy classes.

The selection of results which discussed in the book is mainly based on author's interests. We did not try to give a complete survey. Nevertheless, we tried to give complete proofs almost to all presented theorems.

I am grateful to Prof. N. S. Bakhvalov who posed me a series of optimal recovery problems. I am also deeply indebted to Prof. V. M. Tikhomirov for many fruitful discussions.

Moscow, 1999 K. Yu. Osipenko

Introduction

What does it mean to solve a problem by an optimal way? Assume that we have a set of methods (or algorithms) $\mathcal{M}$ for the solution of a problem p. Each method uses some information Ip about the problem p. To compare different methods of the solution with each method we associate a number which indicate the error of the solution. We denote this number by $e(p, I, m)$.

Usually we need to solve not one problem but several problems of the same type. Denote the set of such problems by $\mathcal{P}$. If we wish to find a method for which the error of the solution will be as small as possible for all our problems, we obtain the following extremal problem: to find the value

$$e(\mathcal{P}, I, \mathcal{M}) := \inf_{m \in \mathcal{M}} \sup_{p \in \mathcal{P}} e(p, I, m)$$

and the method for which the infimum is attained. Every such method we call an *optimal method.*

If $e(\mathcal{P}, I, \mathcal{M})$ is not sufficiently small we may tray to find another type of information which gives a better error. In other words, we can consider the problem:

$$\inf_{I \in \mathcal{I}} e(\mathcal{P}, I, \mathcal{M})$$

where $\mathcal{I}$ is some set of information operators. Every information for which this infimum is attained we call an *optimal information* .

Let us consider some examples.

Example 1 (optimal interpolation). Let W be some class of functions defined on a domain D. Denote by p_f the problem of finding $f(t)$, $t \in D$, for a function $f \in W$. Put

$$Ip_f = (f(t_1), \dots, f(t_n)), \quad t_j \in D.$$

Let $\mathcal{M}$ be the set of all functions $m\colon C^n \to C$ (or $m\colon R^n \to R$ in the real case). Then we have the following problem

$$e(t, W, I) := \inf_{m\colon C^n \to C} \sup_{f \in W} |f(t) - m(Ip_f)|.$$

This problem is called the problem of *optimal recovery* of $f \in W$ at a point t from the information about the values $f(t_1), \dots, f(t_n)$.

Example 2 (optimal integration). Let p_f be the problem of finding the integral

$$\int_a^b f(t)\,dt$$

for the function $f \in W$. With the same Ip_f and $\mathcal{M}$ we obtain the problem of optimal integration of function f on the class W from the values at a fixed system of nodes

$$e(W, I) := \inf_{m\colon C^n \to C} \sup_{f \in W} \left| \int_a^b f(t)\,dt - m(Ip_f) \right|.$$

If we consider the set $\mathcal{M}_0$ containing only linear functions m, then we obtain the problem of optimal quadrature formula. We can also minimize this value over all nodes from $[a, b]$. Thus we obtain the problem of optimal nodes for optimal method of integration.

In general, a similar problem of optimal recovery may be considered for an arbitrary linear functional Lf. In particular, for $Lf = f'(t)$ we obtain the problem of optimal numerical differentiation.

Example 3 (n-widths). Let X be a normed linear space, p_x be the problem of finding $x \in W \subset X$ by means of finite-dimensional subspaces of X. Here $Ip_x = x$, $\mathcal{M}$ is the set of all mappings $m\colon X \to Y_n$, $\dim Y_n \le n$. Putting $e(p_x, I, m) = \|x - m(x)\|_X$ we obtain the problem of finding the value

$$d_n(W, X) := \inf_{Y_n} \inf_{y \in Y_n} \sup_{x \in W} \|x - y\|_X$$

which is known as the Kolmogorov n-width. When $\mathcal{M}$ is the set of all linear mappings the appropriate value is called the linear n-width.

In the present book we deal with the considered problems and some others for classes of analytic functions, mainly the Hardy classes. In Chapter 1 we give general settings and general results on optimal recovery problems. Chapter 2 deals with optimal recovery of functions from the Hardy classes at a fixed point. We also consider the problem of optimal nodes and the case of several variables. In Chapter 3 we study the problems of optimal numerical differentiation for functions from the same classes. We also consider the case when the information about functions is given with an error. In particular, we obtain the Kolmogorov inequalities for analytic functions.

Chapter 4 is devoted to n-widths of various classes of analytic functions. We give a preference to exact results but there are some asymptotic estimates, too. In Chapter 5 we consider best and optimal quadrature formulas for the Hardy classes H_p. Most of the considered extremal problems for analytic functions are closely connected with elliptic functions. For the convenience we collected some of the needed notions and results in Appendix. GK. Yu. Osipenko Optimal Recovery Settings GK. Yu. Osipenko Optimal Recovery Settings

Chapter 1

Optimal Recovery Settings and General Results

In this chapter we give general settings and prove general results on optimal recovery problems. In Section 1.1 we pose the general problem of optimal recovery of operators from inaccurate information. In Section 1.2 we generalize this problem for optimal recovery of multi-valued mappings and obtain necessary and sufficient conditions for the existence of linear optimal methods of recovery. Section 1.3 is devoted to optimal recovery of linear functionals. In this section we prove the main theorems (Theorems 1.9 and 1.10) which we almost always apply to find optimal methods. In Section 1.4 the obtained results are used for the problem of optimal recovery of functions and their derivatives on the basis of information about the Fourier coefficients.

1.1 A General Setting

By a *multi-valued* mapping $F\colon W \to V$ we shall mean a map which associates with every $x \in W$ a nonempty set $F(x) \subset V$. The set

$$\operatorname{gr} F := \{\, (x, y) : x \in W,\ y \in F(x) \,\}$$

is called the *graph* of F.

A general problem of optimal recovery of operators from inaccurate information may be posed in the following way. Let X be a linear space, Z a normed linear space and $L\colon X \to Z$ a linear operator. We wish to recover the values of L on a set $W \subset X$ on the basis of the values of multi-valued mapping $F\colon W \to V$.

The multivalence of F is simulating inaccurate information about elements from W. If F is a single-valued mapping, then we say about optimal recovery problem from exact data. We shall often consider the case when $F(x) = Ix + U$, where $I\colon X \to Y$ is a linear operator, Y is a linear space, and U is a fixed set from Y. The operator I is called an *information operator*. For example, if Y is a normed linear space, then we can set $U := \{y \in Y : \|y\| \le \delta\}$. In this case we say that the information operator is given with an *error* δ.

By a *method* (or *algoritm*) *of recovery* of operator L we mean any map $\varphi\colon F(W) \to Z$. Thus we have the diagram

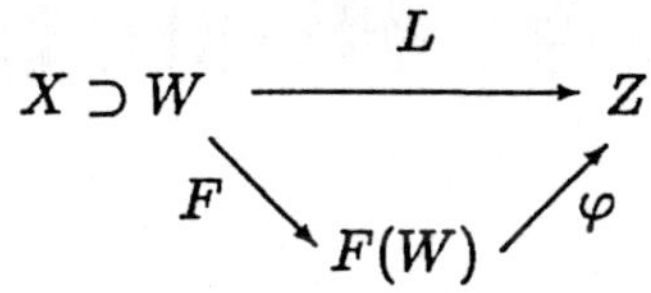

The *error of a method* φ is defined as follows

$$e(L, F, \varphi) := \sup_{(x,y)\in \operatorname{gr} F} \|Lx - \varphi(y)\|.$$

Suppose we have a set of methods $\mathcal{S}$. Put

$$e(L, F, \mathcal{S}) := \inf_{\varphi\in\mathcal{S}} e(L, F, \varphi). \tag{1.1}$$

Assume that we wish to recover a family of operators $\mathcal{L}$ and we can choose mappings F from a set $\mathcal{F}$. Put

$$e(\mathcal{L}, \mathcal{F}, \mathcal{S}) := \inf_{F\in\mathcal{F}} \sup_{L\in\mathcal{L}} e(L, F, \mathcal{S}). \tag{1.2}$$

Let us consider some examples of problems (1.1) and (1.2). Denote by $\mathcal{E}$ the set of all possible methods $\varphi\colon F(W) \to Z$. Suppose that $\mathcal{S} = \mathcal{E}$ and W is a class of real-valued or complex-valued functions defined on a set G. Set $\tau := (t_1, \dots, t_n)$, $t_j \in G$,

$$I_\tau x := \{\, x(t_1), \dots, x(t_n) \,\},$$

and $F = F_\tau := I_\tau + \delta B_n$, where B_n is the unit ball of R^n or C^n in which some norm is introduced. Then for $Lx = L_t x := x(t)$, $t \in G$, the problem (1.1) is the one of optimal recovery of function $x \in W$ at the point t by the values of this function given with the error δ at the system of points τ.

If $E \subset G$ and

$$\mathcal{L} = \{ L_t : t \in E \}, \qquad \mathcal{F} = \{ F_\tau : \tau \in E^n \},$$

then (1.2) is the problem of optimal recovery of functions from W on the set E by the values given with the error δ.

If

$$Lx = \int_E x(t)\, dt,$$

then (1.1) is the problem of best method of integration using approximate values of x at a fixed system of nodes and (1.2) with $\mathcal{L} = \{L\}$ is the problem of optimal method of integration (or the problem of optimal choice of nodes).

The proposed scheme includes also a number of other well-known problems of approximation theory such as optimal recovery of derivatives, optimal recovery from the Fourier coefficients, approximation of unbounded operators by bounded operators, n-widths, and so on.

Let us consider more explicitly the settings of problems with n-widths. Denote by $\mathcal{E}_n$ the set of all operators of $F(W)$ into Z of rank at most n. Denote By $\mathcal{E}_n^\lambda$ the set of all linear operators of rank at most n which map any spaces containing $F(W)$ into Z. Put

$$d_n(L, F) := e(L, F, \mathcal{E}_n), \qquad \lambda_n(L, F) := e(L, F, \mathcal{E}_n^\lambda).$$

If $F^{-1}(0) \neq \emptyset$ (the mapping F^{-1} is the multi-valued mapping defined by $F^{-1}(y) := \{x \in W : (x, y) \in \operatorname{gr} F\}$), then we have

$$\lambda_n(L, F) \geq \inf_{\varphi \in \mathcal{E}_n^\lambda} \sup_{\substack{(x,y) \in \operatorname{gr} F \\ \varphi(y) = 0}} \|Lx\| =: d^n(L, F).$$

Set $I_\delta := Ix + \delta BY$, where $I\colon W \to Y$ is a linear operator, $\delta \geq 0$, and

$$BY := \{ y \in Y : \|y\| \leq 1 \}.$$

Then for $F = I_\delta$ and $L = Id$ (Id is an identical mapping) we can define the values

$$d_n(W, X, I, \delta) := d_n(Id, I_\delta), \qquad \lambda_n(W, X, I, \delta) := \lambda_n(Id, I_\delta),$$
$$d^n(W, X, I, \delta) := d^n(Id, I_\delta).$$

For $\delta = 0$ and $L = Id$ the values

$$\begin{gathered} d_n(W, X) := d_n(W, X, Id, 0), \qquad \lambda_n(W, X) := \lambda_n(W, X, Id, 0), \\ d^n(W, X) := d^n(W, X, Id, 0) \end{gathered} \tag{1.3}$$

are well known in approximation theory. They are called the Kolmogorov, linear, and Gel'fand n-widths, respectively.

1.2 Optimal Recovery of Multi-Valued Mappings

The problem (1.1) may be reduced to a more general problem relating to optimal recovery of multi-valued mappings. Taking $\Phi(y) := L(F^{-1}(y))$ it can be easily shown that

$$e(L, F, \varphi) = \sup_{(y,z)\in \operatorname{gr} \Phi} \|z - \varphi(y)\|.$$

Let us formulate now a general problem of optimal recovery of multi-valued mapping by single-valued mappings. Let $\Phi\colon A \to Z$ be a multi-valued mapping, Z a normed linear space, and $\varphi\colon A \to Z$ a single-valued mapping (a method of recovery of Φ). Set

$$E(\Phi, \varphi) := \sup_{(y,z)\in \operatorname{gr} \Phi} \|z - \varphi(y)\|.$$

Suppose that $\mathcal{S}$ is a set of single-valued mappings $\varphi\colon A \to Z$. The value

$$E(\Phi, \mathcal{S}) := \inf_{\varphi\in\mathcal{S}} E(\Phi, \varphi). \tag{1.4}$$

is called the *intrinsic error* of the problem. A method for which the infimum in (1.4) is attained is said to be an *optimal method of recovery* for a given $\mathcal{S}$.

Set

$$E(\Phi) := E(\Phi, \mathcal{E}),$$

where $\mathcal{E}$ is the set of all possible mappings of A into Z.

The value

$$R(\Phi) := \sup_{y\in A} \inf_{c\in Z} \sup_{z\in\Phi(y)} \|z - c\|$$

which is called the *radius of multivalence* of Φ is closely connected to the problem (1.4) for $S = \mathcal{E}$. The radius of multivalence is the maximal Chebyshev radius of the sets $\Phi(y)$ when $y \in A$.

Lemma 1.1.

$$E(\Phi) = R(\Phi).$$

Proof. Let $\varepsilon > 0$. For every $y \in A$ we denote by $\varphi_\varepsilon(y)$ any element of the set

$$\{\, d \in Z : \sup_{z\in\Phi(y)} \|z - d\| \le \inf_{c\in Z} \sup_{z\in\Phi(y)} \|z - c\| + \varepsilon \,\}.$$

Then

$$E(\Phi) \le E(\Phi, \varphi_\varepsilon) = \sup_{y\in A} \sup_{z\in\Phi(y)} \|z - \varphi_\varepsilon(y)\| \le R(\Phi) + \varepsilon.$$

Since $\varepsilon > 0$ is arbitrary, we have

$$E(\Phi) \le R(\Phi).$$

The inequality

$$E(\Phi) \ge R(\Phi)$$

evidently follows from the definition. □

In what follows we shall consider nonempty sets $A \subset X$, where X are linear spaces over the field $K = R$ or C. If a field is not explicitly indicated in some statement, then it is related to both cases.

Denote by X' the space algebraically dual to X, that is the space of all linear functionals on X. The annihilator of a linear subspace $X_0 \subset X$ is defined as

$$X_0^\perp := \{\, x' \in X' : \langle x', x\rangle = 0 \;\; \forall x \in X_0 \,\}.$$

Denote by $\operatorname{co} A$ and $\operatorname{bco} A$ the *convex* and *convex balanced hulls* of A:

$$\operatorname{co} A := \left\{ x : x = \sum_{j=1}^{n} \lambda_j x_j, \quad x_j \in A, \quad \sum_{j=1}^{n} \lambda_j = 1 \right\};$$

$$\operatorname{bco} A := \left\{ x : x = \sum_{j=1}^{n} \lambda_j x_j, \quad x_j \in A, \quad \sum_{j=1}^{n} |\lambda_j| \le 1 \right\}.$$

Lemma 1.2. *Let X be a linear space, $X_0 \subset X$ a linear subspace, $A \subset X$, and $x_0' \in X'$. Then*

$$\inf_{x'\in X_0^\perp} \sup_{x\in A} |\langle x_0' - x', x\rangle| = \sup_{x\in \operatorname{bco} A \cap X_0} |\langle x_0', x\rangle|$$

and the infimum is attained.

Proof. 1. The lower bound. Let $x \in \operatorname{bco} A$ and $x' \in X'$. Then $x = \sum_{j=1}^{n} \lambda_j x_j$, $x_j \in A$, $\sum_{j=1}^{n} |\lambda_j| \le 1$, and

$$|\langle x', x\rangle| = |\sum_{j=1}^{n} \lambda_j \langle x', x_j\rangle| \le \max_{1\le j\le n} |\langle x', x_j\rangle| \le \sup_{x\in A} |\langle x', x\rangle|.$$

Hence

$$\sup_{x\in A} |\langle x', x\rangle| \ge \sup_{x\in \operatorname{bco} A} |\langle x', x\rangle|.$$

Since the reverse inequality is evident we have for all $x' \in X'$

$$\sup_{x \in A} |\langle x', x\rangle| = \sup_{x \in \operatorname{bco} A} |\langle x', x\rangle|.$$

Let $x' \in X_0^{\perp}$. Then

$$\begin{aligned}\sup_{x \in A} |\langle x_0' - x', x\rangle| = \sup_{x \in \operatorname{bco} A} |\langle x_0' - x', x\rangle| &\geq \sup_{x \in \operatorname{bco} A \cap X_0} |\langle x_0' - x', x\rangle| \\ &= \sup_{x \in \operatorname{bco} A \cap X_0} |\langle x_0', x\rangle| =: \rho.\end{aligned}$$

2. The upper bound. For $\rho = \infty$ the assertion of lemma follows from 1. Let $\rho < \infty$. Denote by $\mu(x) := \inf\{t > 0 : x \in t \operatorname{bco} A\}$ the Minkowski functional of $\operatorname{bco} A$. Set

$$p(x) := \begin{cases} \infty, & \mu(x) = \infty, \\ \rho\mu(x), & \mu(x) < \infty. \end{cases}$$

Let us show that for all $x \in X_0$

$$|\langle x_0', x\rangle| \leq p(x). \tag{1.5}$$

For $\mu(x) = \infty$ (1.5) is evident. If $x \in X_0$ and $\mu(x) < \infty$, then since $\mu(x_0) < 1$, for all $\varepsilon > 0$

$$x_0 := \frac{x}{\mu(x) + \varepsilon} \in \operatorname{bco} A \cap X_0.$$

Consequently,

$$|\langle x_0', x\rangle| = |\langle x_0', x_0\rangle|(\mu(x) + \varepsilon) \leq \rho(\mu(x) + \varepsilon).$$

Since $\varepsilon > 0$ is arbitrary, (1.5) is proved. From the Hahn-Banach theorem there exists an extension of x_0' to $x' \in X'$ such that for all $x \in X$

$$|\langle x', x\rangle| \leq p(x).$$

Put $\hat{x}' := x_0' - x' \in X_0^{\perp}$. Then for all $x \in \operatorname{bco} A$

$$|\langle x_0' - \hat{x}', x\rangle| = |\langle x', x\rangle| \leq p(x) = \rho\mu(x) \leq \rho.$$

Hence

$$\sup_{x \in A} |\langle x_0' - \hat{x}', x\rangle| \leq \rho.$$

□

The functional $a(x)$ is called *affine* if $a(x) = \langle x', x\rangle + c$, where $x' \in X$ and $c \in K$. Denote by X^a the set of all affine functionals. We shall be interested in conditions when there exist linear or affine methods among all optimal methods of recovery.

Define the multi-valued mappings $\operatorname{co}\Phi$ and $\operatorname{bco}\Phi$ as follows

$$\operatorname{gr}\operatorname{co}\Phi := \operatorname{co}\operatorname{gr}\Phi, \qquad \operatorname{gr}\operatorname{bco}\Phi := \operatorname{bco}\operatorname{gr}\Phi.$$

Theorem 1.3. 1. *For the existence of $x' \in X'$ such that*

$$E(\Phi, x') = E(\Phi),$$

it is necessary and sufficient that

$$R(\Phi) = R(\operatorname{bco}\Phi). \tag{1.6}$$

Moreover,

$$E(\Phi) = \sup_{c\in \operatorname{bco}\Phi(0)} |c|.$$

2. *If $K = R$ then for the existence of $a \in X^a$ such that*

$$E(\Phi, a) = E(\Phi), \tag{1.7}$$

it is necessary and sufficient that

$$R(\Phi) = R(\operatorname{co}\Phi). \tag{1.8}$$

Proof. 1. Set $\widetilde{X} := X \times K$, $\widetilde{X}_0 := \{0\} \times K$. Define $\tilde{x}'_0 \in \widetilde{X}'$ by the equality

$$\langle \tilde{x}'_0, (x, c)\rangle := c, \qquad x \in X, \quad c \in K.$$

For any $\tilde{x}' \in \widetilde{X}'$ we have

$$\langle \tilde{x}', (x, c)\rangle = \langle x', x\rangle + \lambda c,$$

where $x' \in X'$ and $\lambda \in K$. Therefore $\tilde{x} \in \widetilde{X}_0^{\perp}$ if and only if $\langle \tilde{x}', (x, c)\rangle = \langle x', x\rangle$. From Lemma 1.2 we have

$$\inf_{x'\in X'} E(\Phi, x') = \inf_{\tilde{x}'\in \widetilde{X}_0^{\perp}} \sup_{(x,c)\in \operatorname{gr}\Phi} |\langle \tilde{x}'_0 - \tilde{x}', (x, c)\rangle|$$

$$= \sup_{(x,c)\in \operatorname{bco}\operatorname{gr}\Phi\cap\widetilde{X}_0} |\langle \tilde{x}'_0, (x, c)\rangle| = \sup_{c\in \operatorname{bco}\Phi(0)} |c|.$$

Moreover, the infimum is attained. In view of Lemma 1.1 and since $\operatorname{bco}\Phi(0)$ is balanced we have

$$R(\operatorname{bco}\Phi) = E(\operatorname{bco}\Phi) \le \inf_{x'\in X'} E(\Phi, x') = \sup_{c\in \operatorname{bco}\Phi(0)} |c| \le R(\operatorname{bco}\Phi).$$

Hence

$$\sup_{c\in \operatorname{bco}\Phi(0)} |c| = R(\operatorname{bco}\Phi) = \inf_{x'\in X'} E(\Phi, x').$$

Taking into account Lemma 1.1 we obtain that (1.6) is equivalent to the equality

$$E(\Phi) = \inf_{x'\in X'} E(\Phi, x')$$

in which the infimum is obtained.

2. Let $K = R$. Define the multi-valued mapping Φ_0 by the equality

$$\operatorname{gr}\Phi_0 := \operatorname{gr}\operatorname{co}\Phi - \operatorname{gr}\operatorname{co}\Phi := \{(x_1, c_1) - (x_2, c_2) : (x_j, c_j) \in \operatorname{gr}\operatorname{co}\Phi,\ j = 1, 2\}.$$

Since $\operatorname{bco}\Phi_0 = \Phi_0$, we have proved that there exists $x' \in X'$ for which

$$E(\Phi, x') = \sup_{c\in\Phi_0(0)} c = \sup_{(x_j,c_j)\in \operatorname{gr}\operatorname{co}\Phi,\ j=1,2} (c_1 - c_2) = 2R(\operatorname{co}\Phi).$$

Thus for all $(x_j, c_j) \in \operatorname{gr}\operatorname{co}\Phi$, $j = 1, 2$,

$$|c_1 - c_2 - \langle x', x_1 - x_2\rangle| \le 2R(\operatorname{co}\Phi).$$

Consequently,

$$c_1 - \langle x', x_1\rangle - R(\operatorname{co}\Phi) \le c_2 - \langle x', x_2\rangle + R(\operatorname{co}\Phi).$$

It is easy to see that there exists $c_0 \in R$ such that

$$c_1 - \langle x', x_1\rangle - R(\operatorname{co}\Phi) \le c_0 \le c_2 - \langle x', x_2\rangle + R(\operatorname{co}\Phi).$$

It means that

$$\sup_{(x,c)\in \operatorname{gr}\operatorname{co}\Phi} |c - \langle x', x\rangle - c_0| \le R(\operatorname{co}\Phi).$$

Using Lemma 1.1 we have for $a(x) := \langle x', x\rangle + c_0$

$$R(\operatorname{co}\Phi) = E(\operatorname{co}\Phi) \le E(\operatorname{co}\Phi, a) \le R(\operatorname{co}\Phi).$$

Hence

$$E(\operatorname{co}\Phi, a) = R(\operatorname{co}\Phi).$$

It is easily seen that

$$E(\Phi, a) = E(\operatorname{co}\Phi, a).$$

Thus there exists $a \in X^a$ for which

$$E(\Phi, a) = R(\operatorname{co}\Phi).$$

By virtue of Lemma 1.1 this proves the equivalence of conditions (1.7) and (1.8). □

We return to the problem (1.1) for the case when $L = x' \in X'$ (i.e., $Z = K$) and S coincides with the set $\mathcal{E}$ of all functionals on Y. Set

$$e(x', F) := e(x', F, \mathcal{E}). \tag{1.9}$$

As it was mentioned above, this problem reduces to the problem of optimal recovery of the multi-valued mapping $\Phi = x' \circ F^{-1}$. Moreover,

$$e(x', F) = E(x' \circ F^{-1}).$$

Let us introduce the value

$$r(x', F) := \sup_{y \in F(W)} \inf_{c \in K} \sup_{x \in F^{-1}(y)} |\langle x', x\rangle - c|,$$

which is called the *information radius*. It is easily seen that

$$r(x', F) = R(x' \circ F^{-1}).$$

Since

$$\operatorname{co}(x' \circ F^{-1}) = x' \circ (\operatorname{co} F)^{-1}, \qquad \operatorname{bco}(x' \circ F^{-1}) = x' \circ (\operatorname{bco} F)^{-1},$$

from Theorem 1.3 it follows

Theorem 1.4. 1. *For the existence of $y' \in Y'$ such that*

$$e(x', F, y') = e(x', F),$$

it is necessary and sufficient that

$$r(x', F) = r(x', \operatorname{bco} F).$$

Moreover,

$$e(x', F) = \sup_{x \in \operatorname{bco} F^{-1}(0)} |\langle x', x\rangle|.$$

2. *If $K = R$ then for the existence of $a \in Y^a$ such that*

$$e(x', F, a) = e(x', F),$$

it is necessary and sufficient that

$$r(x', F) = r(x', \operatorname{co} F).$$

We call a multi-valued mapping F *convex* (*convex balanced*), if

$$\operatorname{co} F = F, \qquad (\operatorname{bco} F = F).$$

Corollary 1.5. *If F is a convex (convex balanced) multi-valued mapping, then among optimal methods there exists an affine (linear) method.*

1.3 Optimal Recovery of Linear Functionals

Let us consider the problem (1.9) for the multi-valued mapping

$$F(x) = Ix + U, \tag{1.10}$$

where $I\colon W \to Y$ is a linear operator and $U \subset Y$ is some set. In this case we set

$$e(x', I, W, U) := e(x', F).$$

Theorem 1.6. *Let $x' \in X'$ and W, U be convex balanced sets. Then among optimal methods there exists a linear method and*

$$\begin{aligned} e(x', I, W, U) &= \sup_{\substack{x \in W \\ Ix \in U}} |\langle x', x\rangle| \\ &= \inf_{y' \in Y'} \left(\sup_{x \in W} |\langle x', x\rangle - \langle y', Ix\rangle| + \sup_{y \in U} |\langle y', y\rangle| \right). \end{aligned} \tag{1.11}$$

Moreover, $y' \in Y'$ is an optimal method of recovery if and only if it brings the lower bound in (1.11).

Proof. The multi-valued mapping F defined by (1.10) is convex and balanced. From Corollary 1.5 it follows the existence of linear optimal method of recovery. We have

$$e(x', I, W, U) = \sup_{x \in F^{-1}(0)} |\langle x', x\rangle| = \sup_{\substack{x \in W \\ Ix \in U}} |\langle x', x\rangle|.$$

Since W and U are convex balanced sets, for every $y' \in Y'$

$$\sup_{\substack{x\in W\\ y\in U}} |\langle x', x\rangle - \langle y', Ix + y\rangle| = \sup_{x\in W} |\langle x', x\rangle - \langle y', Ix\rangle| + \sup_{y\in U} |\langle y', y\rangle|.$$

Therefore

$$\begin{aligned} e(x', I, W, U) &= \inf_{y'\in Y} \sup_{\substack{x\in W\\ y\in U}} |\langle x', x\rangle - \langle y', Ix + y\rangle| \\ &= \inf_{y'\in Y'} \left(\sup_{x\in W} |\langle x', x\rangle - \langle y', Ix\rangle| + \sup_{y\in U} |\langle y', y\rangle| \right). \end{aligned}$$

□

An element $x_0 \in W$ is said to be *extremal*, if $Ix_0 \in U$ and

$$\sup_{\substack{x\in W\\ Ix\in U}} |\langle x', x\rangle| = |\langle x', x_0\rangle|.$$

If W and U are balanced sets, then an extremal element is defined up to factor λ, $|\lambda| = 1$. In this case let us normalize an extremal element by the condition

$$\sup_{\substack{x\in W\\ Ix\in U}} |\langle x', x\rangle| = \langle x', x_0\rangle.$$

Theorem 1.7. *Let W and U be convex balanced sets. Then $x_0 \in W$ is an extremal element and $y'_0 \in Y'$ is an optimal method if and only if*

(i) $\sup_{x\in W} |\langle x', x\rangle - \langle y'_0, Ix\rangle| = \langle x', x_0\rangle - \langle y'_0, Ix_0\rangle$,

(ii) $\sup_{y\in U} |\langle y'_0, y\rangle| = \langle y'_0, Ix_0\rangle$,

(iii) $Ix_0 \in U$.

Proof. Let x_0 be an extremal element and $y'_0 \in Y'$ be an optimal method of recovery. Taking into account (1.11), we have

$$\begin{aligned} \langle x', x_0\rangle &\le |\langle x', x_0\rangle - \langle y'_0, Ix_0\rangle| + |\langle y'_0, Ix_0\rangle| \\ &\le \sup_{x\in W} |\langle x', x\rangle - \langle y'_0, Ix\rangle| + \sup_{y\in U} |\langle y'_0, y\rangle| = \sup_{\substack{x\in W\\ Ix\in U}} |\langle x', x\rangle|. \end{aligned} \tag{1.12}$$

Since x_0 is an extremal element, there are equalities in (1.12). Consequently conditions (i)–(iii) are fulfilled.

In the case when (i)–(iii) are satisfied we have

$$\sup_{\substack{x\in W\\ Ix\in U}} |\langle x', x\rangle| \le \sup_{x\in W} |\langle x', x\rangle - \langle y_0', Ix\rangle| + \sup_{y\in U} |\langle y_0', y\rangle|$$

$$= \langle x', x\rangle - \langle y_0', Ix\rangle + \langle y_0', Ix_0\rangle = \langle x', x_0\rangle.$$

From here it follows that x_0 is an extremal element and that the lower bound in (1.11) is attained for y_0'. From Theorem 1.6 it follows that y_0' is an optimal method of recovery. □

The most usual case for optimal recovery problems with inaccurate information is the case when Y is a normed linear space and $U = U_\delta := \{y \in Y : \|y\| \le \delta\}$, $\delta \ge 0$. Set

$$e(x', I, W, \delta) := e(x', I, W, U_\delta).$$

Denote by Y^* the set of all continuous linear functionals on Y. From Theorem 1.7 we have

Corollary 1.8. *Let W be a convex balanced set and $U = U_\delta$. Then $x_0 \in W$ is an extremal element and $y_0^* \in Y^*$ is an optimal method of recovery if and only if*

(i) $\sup_{x\in W} |\langle x', x\rangle - \langle y_0^*, Ix\rangle| = \langle x', x_0\rangle - \langle y_0^*, Ix_0\rangle,$

(ii) $\langle y_0^*, Ix_0\rangle = \delta\|y_0^*\|,$

(iii) $\|Ix_0\| \le \delta.$

Let S be a nonempty set, Σ a σ-algebra of subsets of S, and μ a nonnegative σ-additive measure on Σ. Denote by $L_p(S, \Sigma, \mu)$ (or briefly $L_p(S)$) the set of all σ-measurable functions with values in $K = R$ or C for which

$$\|x\|_p := \left(\int_S |x(s)|^p \, d\mu\right)^{1/p} < \infty, \quad 1 \le p < \infty,$$

$$\|x\|_\infty := \operatorname*{ess\,sup}_{s\in S} |x(s)| < \infty, \quad p = \infty.$$

In particular, when $S = \{1, 2, \dots\}$ and $\mu(\{j\}) = \mu_j > 0$ the space $L_p(S)$ coincides with the space $l_p(\mu)$ which is the set of vectors $x = (x_1, x_2, \dots)$

such that

$$\|x\|_p := \left(\sum_{j=1}^{\infty} \mu_j |x_j|^p\right)^{1/p} < \infty, \quad 1 \le p < \infty,$$
$$\|x\|_\infty := \sup_j |x_j| < \infty, \quad p = \infty.$$

For $\mu_j \equiv 1$ the corresponding space we denote by l_p.

Set

$$(x, y)_S := \int_S x(s)\overline{y(s)}\, d\mu$$

and for $a \in K,\ 1 \le p < \infty$

$$a_{(p)} := \begin{cases} a|a|^{p-2}, & a \ne 0, \\ 0, & a = 0. \end{cases}$$

If X is a normed linear space denote by BX the unit ball

$$BX := \{\, x \in X : \|x\| \le 1 \,\}.$$

Let $f \in L_{p'}(S),\ 1/p + 1/p' = 1$ (for $p = 1, \infty$, $p' = \infty, 1$, respectively). Consider the problem of optimal recovery of linear functional $(x, f)_S$ on the set BX_p by values of multi-valued mapping $F(x) := Ix + \delta BY$, $\delta \ge 0$, where X_p is a linear subspace of $L_p(S)$, $I\colon X_p \to Y$ is a linear operator, and Y is a normed linear space. Thus we consider the problem of finding the value

$$e(f, I, BX_p(S), \delta) = \inf_{\varphi\colon Y \to K} \sup_{x \in BX_p(S)} \sup_{\substack{y \in Y \\ \|Ix - y\| \le \delta}} |(x, f)_S - \varphi(y)| \tag{1.13}$$

and optimal method of recovery (that is, a method for which the lower bound in (1.13) is attained). The situation when the problem of optimal recovery is considering not on the unit ball BL_p but on the unit ball of some subspace of $L_p(S)$ is typical for classes of analytic or harmonic functions.

In view of duality obtained in Theorem 1.6 this problem is closely connected with the extremal problem

$$\sup_{\substack{x \in BX_p \\ \|Ix\| \le \delta}} |(x, f)_S|. \tag{1.14}$$

A function x_0 for which the upper bound in (1.14) is attained is called *extremal*. In many cases extremal functions (or their form) in (1.14) are well known. However, we shall be interested not only in the intrinsic error of optimal recovery but in an optimal method, too. The majority of methods which we shall construct can be obtained with the help of the following theorem.

Theorem 1.9. *Let $g \in X_p$, $g \neq 0$, $g_0 := g/\|g\|_p$, $\|Ig_0\| \leq \delta$, $y_0^* \in Y^*$, $\langle y_0, Ig_0\rangle = \delta\|y_0^*\|$, and for all $x \in X_p$*

$$(x,f)_S - \langle y_0^*, Ix\rangle = \begin{cases} \alpha(x, g_{(p)})_S, & 1 \leq p < \infty, \\ (x, \varphi g)_S, & p = \infty, \end{cases} \tag{1.15}$$

where $\alpha > 0$, $\varphi \in L_1(S)$, $\varphi(s) \geq 0$ almost everywhere, and if $p = \infty$, then $|g(s)| = 1$ almost everywhere. Then y_0^* is an optimal method of recovery, g_0 is an extremal function, and

$$e(f, I, BX_p, \delta) = (g_0, f)_S = \begin{cases} \alpha\|g\|_p^{p-1} + \delta\|y_0^*\|, & 1 \leq p < \infty, \\ \|\varphi\|_1 + \delta\|y_0^*\|, & p = \infty. \end{cases}$$

Proof. Let $1 \leq p < \infty$. Then from (1.15) and the Hölder inequality we have

$$\sup_{x \in BX_p} |(x,f)_S - \langle y_0^*, Ix\rangle| = \alpha \sup_{x \in BX_p} |(x, g_{(p)})_S| \leq \alpha\|g_{(p)}\|_{p'} = \alpha\|g\|_p^{p-1}.$$

On the other hand, we obtain

$$\sup_{x \in BX_p} |(x,f)_S - \langle y_0^*, Ix\rangle| \geq \alpha|(g_0, g_{(p)})_S| = \alpha\|g\|_p^{p-1}.$$

Taking into account (1.15), we have

$$(g_0, f)_S = \langle y_0^*, Ig_0\rangle + \alpha(g_0, g_{(p)})_S = \delta\|y_0^*\| + \alpha\|g\|_p^{p-1}.$$

Consequently,

$$\sup_{x \in BX_p} |(x,f)_S - \langle y_0^*, Ix\rangle| = (g_0, f)_S - \langle y_0^*, Ig_0\rangle. \tag{1.16}$$

For $p = \infty$ the equality (1.16) is proved similarly. From Corollary 1.8 it follows that y_0 is an optimal method and g_0 is an extremal function. By Theorem 1.6

$$e(f, I, BX_p, \delta) = (g_0, f)_S.$$

□

In the case when $\delta = 0$ it is more convenient to use the following theorem which proves analogously to the previous proof.

Theorem 1.10. *Let $g \in X_p$, $g \neq 0$, $Ig = 0$, $y_0^* \in Y^*$ and for all $x \in X_p$*

$$(x, f)_S - \langle y_0^*, Ix\rangle = \begin{cases} \alpha(x, g_{(p)})_S, & 1 \leq p < \infty, \\ \alpha(x, \varphi g)_S, & p = \infty, \end{cases} \tag{1.17}$$

where $\alpha \in C$, $\varphi \in L_1(S)$, $\varphi(s) \geq 0$ almost everywhere, and if $p = \infty$, then $|g(s)| = 1$ almost everywhere. Then y_0^* is an optimal method of recovery, $g_0 := g/\|g\|_p$ is an extremal function, and

$$e(f, I, BX_p, 0) = |(g_0, f)_S| = \begin{cases} |\alpha|\|g\|_p^{p-1}, & 1 \leq p < \infty, \\ |\alpha|\|\varphi\|_1, & p = \infty. \end{cases}$$

1.4 Optimal Algorithms Using Fourier Coefficients

Let X be a Hilbert space and $e_1, e_2, \dots$ is a complete orthonormal system in X. Denote by $x_j := (x, e_j)$ the Fourier coefficients of $x \in X$. Consider the problem of optimal recovery of linear functional $< x', x >= (x, f)$, $f \in X$, on the unit ball BX using inaccurate values of the information operator $Ix = (x_1, \dots, x_n)$.

Assume that we know approximate values of Fourier coefficients $(\tilde{x}_1, \dots, \tilde{x}_n)$ such that

$$|x_j - \tilde{x}_j| \leq \delta_j, \quad j = 1, \dots, n.$$

Thus we consider the problem of finding the value

$$e(f, I, BX, \delta) := \inf_{\varphi:\, C^n \to C} \sup_{x \in BX} \sup_{\substack{y \in C^n \\ Ix - y \in U}} |(x, f) - \varphi(y)|, \tag{1.18}$$

where

$$U = \{\, (y_1, \dots, y_n) \in C^n : |y_j| \leq \delta_j,\ j = 1, \dots, n \,\}.$$

Furthermore, we are interested in optimal method of recovery, that is, such method φ for which the lower bound in (1.18) is attained. In what follows we assume for convenience that $|f_j| = |(f, e_j)| > 0$ for all $j \geq 1$.

For $a \in R$ put

$$a_+ := \begin{cases} a, & a > 0, \\ 0, & a \leq 0. \end{cases}$$

Theorem 1.11. *Let $\lambda \in (0, \|f\|]$ be a solution of the equation*

$$\|f\|^2 - \sum_{j=1}^{n} \left(|f_j|^2 - \lambda^2\delta_j^2\right)_+ - \lambda^2 = 0. \tag{1.19}$$

Then

$$(x, f) \approx \sum_{j=1}^{n} \left(1 - \lambda\delta_j|f_j|^{-1}\right)_+ \overline{f}_j\tilde{x}_j \tag{1.20}$$

is an optimal method of recovery and

$$e(f, I, BX, \delta) = \lambda + \sum_{j=1}^{n} \delta_j \left(|f_j| - \lambda\delta_j\right)_+.$$

Proof. First we show that the equation (1.19) has a solution $\lambda \in (0, \|f\|]$. Denote by $F(\lambda)$ the function on the left-hand side of (1.19). This function is continuous for all $\lambda \geq 0$. Moreover,

$$F(0) = \|f\|^2 - \sum_{j=1}^{n} |f_j|^2 > 0.$$

Since $F(\|f\|) < 0$, there exists a $\lambda \in (0, \|f\|]$ which is a solution of (1.19).

For such λ consider the method (1.20). In view of (1.11) we have

$$e(f, I, BX, \delta) \leq \sup_{x \in BX} \left| (x, f) - \sum_{j=1}^{n} \left(1 - \lambda\delta_j|f_j|^{-1}\right)_+ \overline{f}_j x_j \right|$$
$$+ \sum_{j=1}^{n} \delta_j|f_j| \left(1 - \lambda\delta_j|f_j|^{-1}\right)_+ = \sup_{x \in BX} (x, f_\lambda) + \sum_{j=1}^{n} \delta_j \left(|f_j| - \lambda\delta_j\right)_+$$

where

$$(f_\lambda)_j = \begin{cases} f_j, & j \geq n+1, \\ f_j - f_j\left(1 - \lambda\delta_j|f_j|^{-1}\right)_+, & 1 \leq j \leq n. \end{cases}$$

It can be easily shown that

$$\|f_\lambda\|^2 = \|f\|^2 - \sum_{j=1}^{n} \left(|f_j|^2 - \lambda^2\delta_j^2\right)_+ = \lambda^2.$$

Consequently,

$$e(f, I, BX, \delta) \leq \lambda + \sum_{j=1}^{n} \delta_j \left(|f_j| - \lambda\delta_j\right)_+ .$$

Put

$$x_0 := \frac{f_\lambda}{\|f_\lambda\|} = \lambda^{-1} f_\lambda .$$

Let $1 \leq j \leq n$. If $1 - \lambda\delta_j |f_j|^{-1} > 0$, then

$$|(x_0)_j| = \lambda^{-1} |(f_\lambda)_j| = \delta_j .$$

If $1 - \lambda\delta_j |f_j|^{-1} \leq 0$, then

$$|(x_0)_j| = \lambda^{-1} |f_j| \leq \delta_j .$$

Thus $Ix_0 \in U$. From (1.11) we obtain

$$e(f, I, BX, \delta) \geq |(x_0, f)| = \lambda^{-1}\left(\|f\|^2 - \sum_{j=1}^{n} |f_j|^2 \left(1 - \lambda\delta_j |f_j|^{-1}\right)_+\right)$$

$$= \lambda^{-1}\left(\|f\|^2 - \sum_{j=1}^{n} \left(|f_j| + \lambda\delta_j\right)\left(|f_j| - \lambda\delta_j\right)_+ + \lambda \sum_{j=1}^{n} \delta_j \left(|f_j| - \lambda\delta_j\right)_+\right)$$

$$= \lambda + \sum_{j=1}^{n} \delta_j \left(|f_j| - \lambda\delta_j\right)_+ .$$

□

Assume that $\delta_j = \delta\lambda_j$, $\lambda_j > 0$, $j = 1, \ldots, n$, and $\delta \geq 0$.

Theorem 1.12. *Suppose that*

$$|f_1|\lambda_1^{-1} \geq \ldots \geq |f_n|\lambda_n^{-1} .$$

Set

$$\mu_k := \left(\sum_{j=1}^{k} \lambda_j^2 + |f_k|^{-2}\lambda_k^2 \sum_{j=k+1}^{\infty} |f_j|^2\right)^{-1/2}, \quad k = 1, \ldots, n,$$

$\mu_0 := +\infty$, $\mu_{n+1} := 0$, *and* $\Delta_k := [\mu_{k+1}, \mu_k)$, $k = 0, \ldots, n$. *Then for* $\delta \in \Delta_k$, $0 \leq k \leq n$, *the method*

$$(x, f) \approx \sum_{j=1}^{k} \left(1 - \delta \frac{\lambda_j}{|f_j|} \sqrt{\frac{\sum_{j=k+1}^{\infty} |f_j|^2}{1 - \delta^2 \sum_{j=1}^{k} \lambda_j^2}}\right) \overline{f}_j \tilde{x}_j$$

is optimal and

$$e(f, I, BX, \delta) = \sqrt{\sum_{j=k+1}^{\infty} |f_j|^2} \sqrt{1 - \delta^2 \sum_{j=1}^{k} \lambda_j^2} + \delta \sum_{j=1}^{k} \lambda_j |f_j|.$$

Proof. In the considered case the equation (1.19) has the following form:

$$\|f\|^2 - \sum_{j=1}^{n} \left(|f_j|^2 - \lambda^2\delta^2\lambda_j^2\right)_+ - \lambda^2 = 0. \tag{1.21}$$

If $\delta = 0$, then the solution of (1.21) is evident and the theorem follows from Theorem 1.11 immediately. Suppose that $\delta > 0$. Then (1.21) is equivalent to the equation

$$\frac{c^2}{\|f\|^2 - \sum_{j=1}^{n} \left(|f_j|^2 - c^2\lambda_j^2\right)_+} = \delta^2, \tag{1.22}$$

where $c = \lambda\delta$. Denote by $F(c)$ the left-hand side of (1.22). It is easy to show that $F(c)$ is monotonically increasing for $c \geq 0$. Furthermore,

$$F\left(|f_k|\lambda_k^{-1}\right) = \mu_k^2, \quad k = 1, \dots, n.$$

Hence for $\delta \in \Delta_k$, $0 \leq k \leq n$,

$$c = \delta \sqrt{\frac{\sum_{j=k+1}^{\infty} |f_j|^2}{1 - \delta^2 \sum_{j=1}^{k} \lambda_j^2}}$$

is a solution of (1.21). Now the theorem follows from Theorem 1.11. □

Denote by L the linear space of vectors $x = (x_1, x_2, \dots)$, $x_j \in C$, which satisfy the condition

$$\sum_{j=1}^{\infty} \gamma_j |x_j|^2 < \infty,$$

where $\gamma_1 \geq 0$ and $\gamma_j > 0$, $j \geq 2$. Let x' be the linear functional defined by the equality

$$\langle x', x \rangle := \sum_{j=1}^{\infty} x_j \overline{f}_j,$$

where $|f_j| > 0$, $j \geq 2$, and

$$\sum_{j=2}^{\infty} \gamma_j^{-1}|f_j|^2 < \infty.$$

Consider the problem of optimal recovery of the functional x' on the set

$$BL := \left\{ x \in L : \sum_{j=1}^{\infty} \gamma_j |x_j|^2 \leq 1 \right\}$$

from approximate values $(\tilde{x}_1, \dots, \tilde{x}_n)$ such that

$$|x_j - \tilde{x}_j| \leq \delta\lambda_j, \quad \lambda_j > 0, \quad j = 1, \dots, n.$$

Put

$$m := \begin{cases} 1, & \gamma_1 f_1 \neq 0, \\ 2, & \gamma_1 f_1 = 0. \end{cases}$$

Theorem 1.13. *Suppose that*

$$\frac{|f_m|}{\lambda_m \gamma_m} \geq \dots \geq \frac{|f_n|}{\lambda_n \gamma_n}. \tag{1.23}$$

Set

$$\mu_{km} := \left(\sum_{j=m}^{k} \gamma_j \lambda_j^2 + \gamma_k^2 |f_k|^{-2} \lambda_k^2 \sum_{j=k+1}^{\infty} \gamma_j^{-1} |f_j|^2 \right)^{-1/2}, \quad k = m, \dots, n,$$

$\mu_{m-1,m} := +\infty$, $\mu_{n+1,m} := 0$, *and* $\Delta_{km} := [\mu_{k+1,m}, \mu_{km})$, $k = m-1, \dots, n$. *Then for all* $\delta \in \Delta_{km}$, $m - 1 \leq k \leq n$, *the method*

$$\langle x', x \rangle \approx (m-1)\overline{f}_1 \tilde{x}_1 + \sum_{j=m}^{k} \nu_{jm} \overline{f}_j \tilde{x}_j, \tag{1.24}$$

where

$$\nu_{jm} := 1 - \delta \frac{\gamma_j \lambda_j}{|f_j|} \sqrt{\frac{\sum_{j=k+1}^{\infty} \gamma_j^{-1} |f_j|^2}{1 - \delta^2 \sum_{j=m}^{k} \gamma_j \lambda_j^2}},$$

is optimal and

$$e(x', I, BL, \delta) = \sqrt{\sum_{j=k+1}^{\infty} \gamma_j^{-1} |f_j|^2} \sqrt{1 - \delta^2 \sum_{j=m}^{k} \gamma_j \lambda_j^2} + \delta \sum_{j=m}^{k} \lambda_j |f_j|.$$

Proof. Consider the case $m = 1$. Then L is a Hilbert space with the inner product

$$(x, y)_L := \sum_{j=1}^{\infty} \gamma_j x_j \overline{y}_j.$$

The vectors $e_1, e_2, \dots$,

$$(e_j)_s := \begin{cases} 0, & s \neq j, \\ \gamma_j^{-1/2}, & s = j, \end{cases}$$

form a complete orthonormal basis in L. The Fourier coefficients of x in this basis are equal to $(x, e_j) = \sqrt{\gamma_j} x_j$. Using Theorem 1.12 for λ_j and f_j equal $\gamma_j^{1/2}\lambda_j$ and $\gamma_j^{-1/2} f_j$, respectively, we obtain the required statements in the considered case.

Suppose that $\gamma_1 = 0$. Denote by L_0 the space of all vectors $x \in L$ for which $x_1 = 0$. The space L_0 is a Hilbert space with the inner product

$$(x, y)_{L_0} := \sum_{j=2}^{\infty} \gamma_j x_j \overline{y}_j.$$

From Theorem 1.12 it follows that the method

$$\langle x', x \rangle \approx \sum_{j=2}^{k} \nu_{jm} \overline{f}_j \tilde{x}_j$$

is optimal for the set BL_0 (which is the unit ball of L_0), and

$$e(x', I, BL_0, \delta) = \sqrt{\sum_{j=k+1}^{\infty} \gamma_j^{-1} |f_j|^2} \sqrt{1 - \delta^2 \sum_{j=2}^{k} \gamma_j \lambda_j^2} + \delta \sum_{j=2}^{k} \lambda_j |f_j|.$$

From (1.11) we have

$$e(x', I, BL, \delta) = \delta\lambda_1 |f_1| + e(x', I, BL_0, \delta). \tag{1.25}$$

On the other hand, from (1.11) it follows that for the method (1.24)

$$e(x', I, BL, \delta) \leq \sup_{x \in BL_0} \left| \langle x', x \rangle - \sum_{j=2}^{k} \nu_{jm} \overline{f}_j x_j \right| + \delta\lambda_1 |f_1|$$

$$+ \sup_{|z_j| \leq \delta\lambda_j} \left| \sum_{j=2}^{k} \nu_{jm} \overline{f}_j z_j \right| = \delta\lambda_1 |f_1| + e(x', I, BL_0, \delta).$$

In view of (1.25) the method (1.24) is optimal for the set BL.

Now assume that $f_1 = 0$. Then from (1.11)

$$e(x', I, BL, \delta) = e(x', I, BL_0, \delta).$$

Thus in this case it suffices to construct an optimal method for the set BL_0. It can be immediately obtained from Theorem 1.12. □

Let us use Theorem 1.13 to obtain optimal methods of recovery for several concrete classes of functions. Let W be a shift invariant class of sufficiently smooth and 2π-periodic functions. Consider the problem of optimal recovery of $f^{(s)}(t)$, $t \in [0, 2\pi)$, $s \geq 0$, $f \in W$, using information about the Fourier coefficients

$$c_k = \frac{1}{2\pi}\int_0^{2\pi} f(t)e^{-ikt}\,dt, \quad |k| \leq n,$$

given with an error δ in the uniform norm, i.e., by $\tilde{c}_k$ such that

$$|c_k - \tilde{c}_k| \leq \delta, \quad |k| \leq n.$$

Denote by $e_{ns}(W, \delta)$ the intrinsic error for this problem (since W is a shift invariant class, from (1.11) it follows that the intrinsic error does not depend on t).

We shall consider the following classes of periodic analytic functions. The *Hardy space* $\mathcal{H}_2^\beta(T)$ is the set of all 2π-periodic functions analytic in the strip $S_\beta := \{\, z \in C : |\operatorname{Im} z| < \beta \,\}$ which satisfy the condition

$$\|f\|_{\mathcal{H}_2^\beta(T)} := \sup_{0 \leq \eta < \beta} \left(\frac{1}{4\pi} \int_0^{2\pi} \left(|f(t+i\eta)|^2 + |f(t-i\eta)|^2\right) dt \right)^{1/2} < \infty.$$

The *Hardy-Sobolev space* $\mathcal{H}_2^{r,\beta}(T)$ is the set of all 2π-periodic functions analytic in the strip S_β for which $f^{(r)} \in \mathcal{H}_2^\beta(T)$.

The *Bergman space* $\mathcal{A}_2^\beta(T)$ is the set of all 2π-periodic functions analytic in the strip S_β which satisfy the condition

$$\|f\|_{\mathcal{A}_2^\beta(T)} := \left(\frac{1}{4\pi\beta} \int_0^{2\pi} \int_{-\beta}^{\beta} |f(t+i\eta)|^2\, dt d\eta \right)^{1/2} < \infty,$$

and the *Bergman–Sobolev space* $\mathcal{A}_2^{r,\beta}(T)$ is the set of all 2π-periodic functions analytic in the strip S_β for which $f^{(r)} \in \mathcal{A}_2^\beta(T)$.

Functions from $\mathcal{H}_2^\beta(T)$ have finite boundary values almost everywhere and the space $\mathcal{H}_2^\beta(T)$ is a Hilbert space with the inner product

$$(f,g)_{\mathcal{H}_2^\beta(T)} := \frac{1}{4\pi}\int_0^{2\pi}\left(f(t+i\beta)\overline{g(t+i\beta)} + f(t-i\beta)\overline{g(t-i\beta)}\right)dt.$$

The Bergman space $\mathcal{A}_2^\beta(T)$ is also a Hilbert space with the inner product

$$(f,g)_{\mathcal{A}_2^\beta(T)} := \frac{1}{4\pi\beta}\int_0^{2\pi}\int_{-\beta}^{\beta} f(t+i\eta)\overline{g(t+i\eta)}\,dtd\eta.$$

Put

$$H_2^{r,\beta}(T) := \{ f \in \mathcal{H}_2^{r,\beta}(T) : \|f^{(r)}\|_{\mathcal{H}_2^\beta(T)} \le 1 \},$$
$$A_2^{r,\beta}(T) := \{ f \in \mathcal{A}_2^{r,\beta}(T) : \|f^{(r)}\|_{\mathcal{A}_2^\beta(T)} \le 1 \}.$$

It is easy to verify that the functions $e_j(z) := e^{ijz}$, $j = 0, \pm1, \dots$, form a complete orthogonal basis in the spaces $\mathcal{H}_2^\beta(T)$ and $\mathcal{A}_2^\beta(T)$. Moreover,

$$\|e_j\|^2_{\mathcal{H}_2^\beta(T)} = \cosh 2j\beta, \quad j = \pm1, \pm2, \dots ,$$
$$\|e_0\|^2_{\mathcal{A}_2^\beta(T)} = 1, \quad \|e_j\|^2_{\mathcal{A}_2^\beta(T)} = \frac{\sinh 2j\beta}{2j\beta}, \quad j = \pm1, \pm2, \dots .$$

Thus $f \in W = H_2^{r,\beta}(T)$ or $A_2^{r,\beta}(T)$ if and only if

$$f(z) = \sum_{j=-\infty}^{+\infty} c_j e^{ijz}$$

and

$$\sum_{j=-\infty}^{+\infty} |c_j|^2 j^{2r} p_j(W) \le 1$$

where $p_j(W) = \|e_j\|^2_W$.

We introduce the notation

$$\mu_{kr}(W,s) := \left(\sum_{|j|\le k} j^{2r} p_j(W) + k^{4r-2s} p_k^2(W) \sum_{|j|>k} j^{2(s-r)} p_j^{-1}(W)\right)^{-1/2},$$
$$1 \le k \le n,$$

$$\mu_{n+1,r}(W,s) := 0, \quad \mu_{00}(W,s) := \left(\sum_{|j|\geq 0} p_j^{-1}(W)\right)^{-1/2},$$
$$\mu_{0r}(W,s) := +\infty,\ r \geq 1, \quad \Delta_{-1,0}(W,s) := \left[\mu_{00}(W,s), +\infty\right),$$
$$\Delta_{kr}(W,s) := \left[\mu_{k+1,r}(W,s), \mu_{kr}(W,s)\right), \quad 0 \leq k \leq n,\ r \geq 0.$$

Using Theorem 1.13 with $\overline{f}_j = (ij)^s e^{ijt}$, $\lambda_j = 1$, and $\gamma_j = j^{2r} p_j(W)$, we obtain the following result

Theorem 1.14. *Let r and s be nonnegative integers such that $0 \leq s \leq 2r$. Then for all $\delta \in \Delta_{kr}(W,s)$ the method*

$$f^{(s)}(t) \approx \sum_{|j|\leq k} \left(1 - \delta |j|^{2r-s} p_j(W) \sqrt{\frac{\sum_{|j|>k} j^{2(s-r)} p_j^{-1}(W)}{1 - \delta^2 \sum_{|j|\leq k} j^{2r} p_j(W)}}\right) \tilde{c}_j (ij)^s e^{ijt}$$

is optimal for $W = H_2^{r,\beta}(T)$ and $A_2^{r,\beta}(T)$. Moreover,

$$e_{ns}(W,\delta) = \sqrt{\sum_{|j|>k} j^{2(s-r)} p_j^{-1}(W)} \sqrt{1 - \delta^2 \sum_{|j|\leq k} j^{2r} p_j(W)} + \delta \sum_{|j|\leq k} |j|^s.$$

In this theorem we need the condition $0 \leq s \leq 2r$ to satisfy (1.23). Nevertheless, optimal methods of recovery can be obtained for the case $s > 2r$, too. It is sufficient to put the values $j^{s-2r} p_j(W)$ in non-increasing order and use Theorem 1.13.

An optimal method of recovery of $f^{(s)}(t)$, $0 \leq s \leq r-1$, from approximate Fourier coefficients can be also constructed for the Sobolev class $W_2^r(T)$. The class $W_2^r(T)$ may be defined as the set of all functions

$$f(t) = \sum_{j=-\infty}^{+\infty} c_j e^{ijt}$$

for which

$$\sum_{j=-\infty}^{+\infty} j^{2r} |c_j|^2 \leq 1$$

and (since functions from $W_2^r(T)$ are real-valued)

$$c_{-j} = \overline{c}_j, \quad j = 0, 1, \ldots\ . \tag{1.26}$$

The condition (1.26) rather differs this class from the classes $H_2^{r,\beta}(T)$ and $A_2^{r,\beta}(T)$. However, it can be shown that Theorem 1.14 is valid for $W = W_2^r(T)$ and $p_j(W_2^r(T)) \equiv 1$.

1.5 Notes and References

1.1. In 1965 Smolyak posed the problem (1.1) for optimal recovery of a real linear functional by exact values of a finite system of other real linear functionals. Smolyak [1965] proved that in this case for a convex (convex balanced) set there exists a linear (affine) optimal method of recovery. This result was generalized by Osipenko [1976] to the complex case. Marchuk, Osipenko [1975] proved a similar result for the case when the information about finite system of linear functionals is given with an error.

Further generalizations of optimal recovery problems were considered by Micchelli, Rivlin [1977], Arestov [1989], Magaril-Il'yaev, Osipenko [1991] (see also Traub, Woźniakowski [1983], Micchelli, Rivlin [1985], Korneichuk [1987], Tikhomirov [1987], Traub, Wasilkowski, Woźniakowski [1988], Plaskota [1996]). The setting which we consider sometimes is called the worst case setting. For other settings such as average, probabilistic, and randomized settings see Traub, Wasilkowski, Woźniakowski [1988], Plaskota [1996].

1.2. Particular cases of Lemma 1.1 were presented in numerous papers. In a general form it was proved by Arestov [1989]. He also proposed the reduction of optimal recovery problem to the problem of optimal recovery of a multi-valued mapping by a single-valued one. Lemma 1.2 is well known for linear continuous functionals as the duality relation. Since the paper of Nikolskii [1946] duality relations widely explored in extremal and approximation problems (see Ioffe, Tikhomirov [1968], [1974], Khavinson [1963], Korneichuk [1976], [1987], Tikhomirov [1976], [1987]). The criteria of the existence of linear and affine optimal recovery methods (Theorems 1.3 and 1.4) were obtained for the first time by Magaril-Il'yaev, Osipenko [1991].

1.3. The statements which are very close to Theorems 1.6 and 1.7 may be found in Micchelli, Rivlin [1977]. Nevertheless, the second equality of (1.11) and the criterion from Theorem 1.7 for the topological case were appeared previously in Khavinson [1963] in connection with some extremal problems for analytic functions (although problems of optimal recovery were not considered there). Theorems 1.9 and 1.10 are the basic tools for the constructing of optimal recovery methods. With different forms of generality these theorems were proved by Osipenko [1991], Osipenko, Stessin [1991], [1992b].

1.4. The results of this section were obtained by Osipenko [1996]. A similar problem for the estimation of functions in the L_2-norm was considered in Melkman, Micchelli [1979]. The case when the l_2-norm is used to measure the error in the Fourier coefficients was analyzed by Micchelli, Rivlin [1977]. In Boyanov [1976] the problem of optimal recovery of periodic functions from the Sobolev space $\widetilde{W}_q^r$, $1 \leq q \leq \infty$, was solved for the case when the Fourier coefficients are known exactly.

Chapter 2

Optimal Recovery in H_p

This chapter deals with the problem of optimal recovery of functions from the Hardy spaces at a fixed point on the basis of information about function values at a given system of points. In Section 2.1 we construct an optimal method of recovery for the Hardy class H_p. In Section 2.2 we study periodic analogues of the results of Section 2.1. The problem of optimal recovery from the information about function values at countable sets of points is considered in Section 2.3. In particular, we obtain an optimal method of recovery of functions analytic in a strip from the information about their values at the equidistant system of nodes.

Section 2.4 is devoted to optimal nodes of recovery. In Section 2.5 we consider the problem of optimal recovery for the Hardy and Bergman classes in the case of several variables. In particular, we obtain the generalized Schwartz Lemma for these classes (Corollary 2.23 and 2.26).

2.1 Optimal Recovery at a Fixed Point

The *Hardy space* $\mathcal{H}_p$ is the set of all functions analytic in the unit disk $D := \{z \in C : |z| < 1\}$ satisfying the condition

$$\|f\|_{\mathcal{H}_p} := \sup_{0<r<1} \left(\frac{1}{2\pi} \int_0^{2\pi} |f(re^{i\theta})|^p d\theta \right)^{1/p} < \infty, \qquad 1 \le p < \infty,$$

$$\|f\|_{\mathcal{H}_\infty} := \sup_{z \in D} |f(z)| < \infty.$$

It is well known (see, for example, Goluzin [1966], Duren [1970]) that functions from $\mathcal{H}_p$ have boundary values almost everywhere on $|z| = 1$ such that $f(e^{i\theta}) \in L_p(\partial D, \mu)$ where $d\mu(e^{i\theta}) = \dfrac{d\theta}{2\pi}$. Thus the space $\mathcal{H}_p$ can be considered as a subspace of $L_p(\partial D, \mu)$. Moreover, for all $f \in \mathcal{H}_p$ and $\xi \in D$ the Cauchy formula

$$f(\xi) = \frac{1}{2\pi i} \int_{|z|=1} \frac{f(z)}{z - \xi}\, dz$$

holds.

Set

$$H_p := \{ f \in \mathcal{H}_p : \|f\|_{\mathcal{H}_p} \le 1 \}.$$

Consider the problem of optimal recovery of $f \in H_p$ at a point $\xi \in D$ from exact values of the information operator

$$If := \{ f(z_1), \dots, f^{(\nu_1 - 1)}(z_1), \dots, f(z_n), \dots, f^{(\nu_n - 1)}(z_n) \},$$

where $z_1, \dots, z_n$ are distinct points from D. In other words, we shall be interested in the value

$$e(\xi, I, H_p) := \inf_{\varphi:\, C^N \to C} \sup_{f \in H_p} |f(\xi) - \varphi(If)|, \tag{2.1}$$

where $N = \sum_{j=1}^n \nu_j$. Moreover, we wish to find an optimal method of recovery, that is, a method for which the lower bound in (2.1) is attained.

A *Blaschke product of degree m on D* is a function of the form

$$B(z) := \lambda \prod_{j=1}^{m} \frac{z - \alpha_j}{1 - \overline{\alpha}_j z}.$$

Here $|\alpha_j| < 1$, $j = 1, \dots, m$, and $|\lambda| = 1$. It can be easily seen that $|B(z)| \equiv 1$ for $|z| = 1$ and therefore $\overline{B(z)} = B^{-1}(z)$ for all $|z| = 1$. Put

$$W(z) := \prod_{j=1}^{n} \left(\frac{z - z_j}{1 - \overline{z}_j z} \right)^{\nu_j}, \qquad \omega_j(z) := \prod_{\substack{s=1 \\ s \ne j}}^{n} \left(\frac{z - z_s}{1 - \overline{z}_s z} \right)^{\nu_s}.$$

In what follows, all expressions involving p with $p = \infty$ are understood as the limit values as $p \to \infty$.

Theorem 2.1. *For all $1 \le p \le \infty$ the method*

$$f(\xi) \approx \sum_{j=1}^{n} \sum_{\nu=0}^{\nu_j - 1} c_{j\nu}(\xi, p) f^{(\nu)}(z_j), \tag{2.2}$$

where

$$c_{j\nu}(\xi,p)=\frac{W(\xi)(1-|\xi|^2)^{\frac{p-2}{p}}}{\nu!(\nu_j-\nu-1)!}\left(\frac{(1-\overline{z}_jz)^{\nu_j}}{\omega_j(z)(\xi-z)(1-\overline{\xi}z)^{\frac{p-2}{p}}}\right)^{(\nu_j-\nu-1)}\Bigg|_{z=z_j},$$

is optimal for the class H_p. Moreover,

$$e(\xi,I,H_p)=\frac{|W(\xi)|}{(1-|\xi|^2)^{1/p}}.$$

Proof. Put

$$g(z):=\frac{W(z)}{(1-\overline{\xi}z)^{2/p}},\qquad \alpha:=W(\xi)(1-|\xi|^2)^{\frac{p-2}{p}}.$$

It is evident that $g\in\mathcal{H}_p$ for every $\xi\in D$. Let $1\le p<\infty$. Then for all $f\in\mathcal{H}_p$ by the residue theorem we have

$$\alpha\int_{|z|=1}\overline{g(z)}|g(z)|^{p-2}f(z)\,d\mu(z)=\alpha\frac{1}{2\pi i}\int_{|z|=1}\frac{f(z)\,dz}{W(z)(z-\xi)(1-\overline{\xi}z)^{\frac{p-2}{p}}}$$
$$=f(\xi)-\sum_{j=1}^{n}\sum_{\nu=0}^{\nu_j-1}c_{j\nu}(\xi,p)f^{(\nu)}(z_j).$$

Substituting $f(z)=g(z)$ in these equalities, we get

$$\|g\|_{\mathcal{H}_p}=(1-|\xi|^2)^{-1/p}.$$

Now the statement of the theorem follows from Theorem 1.10.

For $p=\infty$, setting $\varphi(z):=(1-\overline{\xi}z)^{-2}$ we have

$$\alpha\int_{|z|=1}\overline{g(z)}|\varphi(z)|f(z)\,d\mu(z)=\alpha\frac{1}{2\pi i}\int_{|z|=1}\frac{f(z)\,dz}{W(z)(z-\xi)(1-\overline{\xi}z)}$$
$$=f(\xi)-\sum_{j=1}^{n}\sum_{\nu=0}^{\nu_j-1}c_{j\nu}(\xi,\infty)f^{(\nu)}(z_j).$$

It remains to use Theorem 1.10. □

Corollary 2.2. *Let $\nu_1=\ldots=\nu_n=1$. Then for all $1\le p\le\infty$ the method*

$$f(\xi)\approx\sum_{j=1}^{n}\frac{\omega_j(\xi)}{\omega_j(z_j)}\frac{1-|z_j|^2}{1-\overline{z}_j\xi}\left(\frac{1-|\xi|^2}{1-z_j\overline{\xi}}\right)^{\frac{p-2}{p}}f(z_j)$$

is optimal for the class H_p.

Consider one more particular case when $n = 1$, $z_1 = 0$, and $\nu_1 = \nu$ ($If = (f(0), \dots, f^{(\nu-1)}(0))$). It is the problem of optimal recovery from Teylor's information. In this case the optimal method (2.2) has the following form:

for $p = \infty$

$$f(\xi) \approx \sum_{j=0}^{\nu-1} \frac{\xi^j}{j!}(1 - |\xi|^{2(\nu-j)}) f^{(j)}(0); \tag{2.3}$$

for $p = 2$

$$f(\xi) \approx \sum_{j=0}^{\nu-1} \frac{\xi^j}{j!} f^{(j)}(0);$$

for $p = 1$

$$f(\xi) \approx \sum_{j=0}^{\nu-2} \frac{\xi^j}{j!} f^{(j)}(0) + \frac{\xi^{\nu-1}}{(\nu-1!)} \frac{1}{1-|\xi|^2} f^{(\nu-1)}(0).$$

Moreover, for all $1 \le p \le \infty$

$$e(\xi, I, H_p) = \frac{|\xi|^\nu}{(1-|\xi|^2)^{1/p}}.$$

2.2 The Periodic Case

Denote by $\mathcal{H}_p^\beta(T)$ the set of all 2π-periodic functions which are analytic in the strip $S_\beta := \{z \in C : |\operatorname{Im} z| < \beta\}$ and satisfy

$$\|f\|_{\mathcal{H}_p^\beta(T)} := \sup_{0 \le \eta < \beta} \left(\frac{1}{4\pi} \int_0^{2\pi} (|f(t+i\eta)|^p + |f(t-i\eta)|^p) \, dt \right)^{1/p} < \infty,$$

$$\|f\|_{\mathcal{H}_\infty^\beta(T)} := \sup_{z \in S_\beta} |f(z)| < \infty.$$

Set

$$H_p^\beta(T) := \{ f \in \mathcal{H}_p^\beta(T) : \|f\|_{\mathcal{H}_p^\beta(T)} \le 1 \}.$$

Consider the problem of optimal recovery of $f \in H_p^\beta(T)$ at some point $t \in T$ from the information operator

$$If := \{ f(t_1), \dots, f^{(\nu_1-1)}(t_1), \dots, f(t_n), \dots, f^{(\nu_n-1)}(t_n) \},$$

where $t_1, \ldots, t_n$ are distinct points from T. Thus we are interested in the intrinsic error of optimal recovery

$$e(t, I, H_p^\beta(T)) := \inf_{\varphi\colon C^N \to C} \sup_{f \in H_p^\beta(T)} |f(t) - \varphi(If)|, \tag{2.4}$$

where $N = \sum_{j=1}^n \nu_j$, and in an optimal method of recovery, that is, such method for which the lower bound in (2.4) is attained.

The solution of this problem is given in terms of elliptic functions. We recall some notions from the elliptic functions theory in Appendix. We denote by $\operatorname{sn}(z,k)$, $\operatorname{cn}(z,k)$, and $\operatorname{dn}(z,k)$ the Jacobi functions (see (A.42)). In what follows we shall indicate the dependence of elliptic functions on the modulus only when it differs from k. Denote by K and K' the complete elliptic integrals of the first kind for the moduli k and $k' := \sqrt{1-k^2}$, respectively (see (A.35)). We shall often deal with the function

$$\kappa(s) := 4e^{-s} \left(\frac{\sum_{m=0}^\infty e^{-2sm(m+1)}}{1 + 2\sum_{m=1}^\infty e^{-2sm^2}} \right)^2, \tag{2.5}$$

(see (A.38)) which coincides with the modulus k such that

$$\frac{\pi K'}{2K} = s.$$

Put

$$W(z) := k^{\frac{N}{2}} \prod_{j=1}^n \operatorname{sn}^{\nu_j} \left(\frac{K}{\pi}(z - t_j) \right).$$

Theorem 2.3. *For all* $1 \le p \le \infty$ *the method*

$$f(t) \approx \sum_{j=1}^n \sum_{\nu=0}^{\nu_j - 1} c_{j\nu}(t,p) f^{(\nu)}(t_j),$$

where

$$c_{j\nu}(t,p) = \frac{K}{\pi} \frac{W(t)}{\nu!(\nu_j - \nu - 1)!} \times \lim_{z \to t_j} \left(\frac{(z-t_j)^{\nu_j} d(z) \operatorname{dn}^{\frac{p-2}{p}} \left(\frac{K}{\pi}(t - z) \right)}{W(z) \operatorname{sn}\left(\frac{K}{\pi}(t-z) \right)} \right)^{(\nu_j - \nu - 1)},$$

$$d(z) = \begin{cases} \operatorname{cn}\left(\frac{K}{\pi}(t-z)\right), & N \text{ even}, \\ \operatorname{dn}\left(\frac{K}{\pi}(t-z)\right), & N \text{ odd}, \end{cases}$$

and $k = \kappa(\beta)$, is an optimal method of recovery for $H_p^\beta(T)$. Moreover,

$$e(t, I, H_p^\beta(T)) = \begin{cases} \left(\frac{2K}{\pi}\right)^{1/p} |W(t)|, & N \text{ even}, \\ \left(\frac{2K}{\pi}\right)^{1/p} \sqrt{k}|W(t)|, & N \text{ odd}. \end{cases}$$

Proof. For $k = \kappa(\beta)$ the function

$$b(z) = \sqrt{k}\operatorname{sn}\frac{K}{\pi}z$$

is analytic in the strip S_β. Moreover, $|b(u + i\beta)| \equiv 1$ for all $u \in R$ (see (A.44)). Thus $\overline{W(z)} = W^{-1}(z)$ for $z \in \partial S_\beta$.

We first consider the case when N is an even number. Since $\operatorname{dn}\frac{K}{\pi}z$ is a 2π-periodic function which does not vanish in the strip S_β and $b(z + 2\pi) = -b(z)$, for all even N

$$g(z) := W(z)\operatorname{dn}^{2/p}\left(\frac{K}{\pi}(t-z)\right) \in \mathcal{H}_\infty^\beta(T).$$

Let $1 \le p < \infty$. Put

$$\alpha := \frac{2K}{\pi}g(t).$$

For $f \in \mathcal{H}_p^\beta(T)$ consider the integral

$$Jf := \frac{\alpha}{4\pi}\int_{\Gamma_0} \overline{g(z)}|g(z)|^{p-2} f(z)\,dz,$$

where $\Gamma_0 := [-i\beta, 2\pi - i\beta] \cup [i\beta, 2\pi + i\beta]$. Using the properties of elliptic functions (see Appendix) we have

$$\overline{\operatorname{dn}\left(\frac{K}{\pi}(u \pm i\beta)\right)} = \pm i\frac{\operatorname{cn}\left(\frac{K}{\pi}(u \pm i\beta)\right)}{\operatorname{sn}\left(\frac{K}{\pi}(u \pm i\beta)\right)}. \tag{2.6}$$

The element of integration in Jf is 2π-periodic. So we can rewrite Jf in the following form

$$Jf = -\frac{KW(t)}{\pi}\frac{1}{2\pi i}\int_{\Gamma_\varepsilon} \frac{\operatorname{cn}\left(\frac{K}{\pi}(t-z)\right)\operatorname{dn}^{\frac{p-2}{p}}\left(\frac{K}{\pi}(t-z)\right)}{W(z)\operatorname{sn}\left(\frac{K}{\pi}(t-z)\right)} f(z)\,dz,$$

where Γ_ε is the boundary of the rectangle $-\varepsilon \le \operatorname{Re} z \le 2\pi - \varepsilon$, $|\operatorname{Im} z| \le \beta$, and ε such that $t, t_1, \dots, t_n$ lie inside this rectangle. By the residue theorem

$$Jf = f(t) - \sum_{j=1}^{n}\sum_{\nu=0}^{\nu_j-1} c_{j\nu}(t,p) f^{(\nu)}(t_j). \tag{2.7}$$

For $f(z) = g(z)$ this equality gives

$$\|g\|_{\mathcal{H}_p^\beta(T)} = \left(\frac{\pi}{2K}\right)^{1/p}.$$

Using Theorem 1.10, we obtain the required statement in the considered case.

If N is odd, then the same arguments should be applied for the function

$$g(z) = \sqrt{k}\operatorname{sn}\left(\frac{K}{\pi}(z-t+\pi)\right) W(z)\operatorname{dn}^{2/p}\left(\frac{K}{\pi}(t-z)\right),$$

using the fact that

$$\operatorname{sn}(u+K) = \frac{\operatorname{cn} u}{\operatorname{dn} u}.$$

For $p = \infty$ consider the integral

$$Jf := \frac{\alpha}{4\pi}\int_{\Gamma_0} \overline{g(z)}|\varphi(z)| f(z)\,dz,$$

where

$$\varphi(z) = \operatorname{dn}^2\left(\frac{K}{\pi}(t-z)\right),$$

and $g(z)$ is defined as above depending on evenness of N. Further similarly to the case $1 \le p < \infty$ we prove (2.7) and use Theorem 1.10. □

Corollary 2.4. *Let $k = \kappa(\beta)$ and $\nu_1 = \ldots = \nu_n = 1$. Then for all $1 \le p \le \infty$ the method*

$$f(t) \approx \frac{K}{\pi} W(t) \sum_{j=1}^{n} \frac{c_{np}\left(\frac{K}{\pi}(t - t_j)\right)}{W'(t_j)} f(t_j),$$

where

$$c_{np}(z) = \begin{cases} \dfrac{\operatorname{cn} z \operatorname{dn}^{\frac{p-2}{p}} z}{\operatorname{sn} z}, & n \text{ even}, \\ \dfrac{\operatorname{dn}^{\frac{2(p-1)}{p}} z}{\operatorname{sn} z}, & n \text{ odd}, \end{cases} \tag{2.8}$$

is an optimal method of recovery for $H_p^\beta(T)$.

2.3 Countable Sets of Nodes

Consider the problem of optimal recovery of functions from H_p knowing their values at countable set of points. Assume that the information operator has the following form

$$If = \left\{ f(z_j), \ldots, f^{(\nu_j - 1)}(z_j) \right\}_{j=-\infty}^{+\infty}, \tag{2.9}$$

where $z_j \in (-1, 1)$. In this case we need an auxiliary lemma to obtain the integral representation (1.17) which is used for the construction of optimal method of recovery.

We recall that by an *infinite Blaschke product for the unit disk D* we mean a function of the form

$$B(z) = \prod_{n=1}^{\infty} -\frac{\overline{z}_n}{|z_n|} \cdot \frac{z - z_n}{1 - \overline{z}_n z}, \tag{2.10}$$

where $z_n \in D$ (for $z_n = 0$ the fraction $-\overline{z}_n/|z_n|$ is replaced by unity and, in accordance with this, we can assume that $\operatorname{sign} 0 = 1$). It is well known (see, for example, Garnett [1981]) that if the $z_n \in D$ satisfy the *Blaschke condition*

$$\sum_{n=1}^{\infty} (1 - |z_n|) < \infty,$$

then the product (2.10) converges in D, $B(z) \in \mathcal{H}_\infty$, and $|B(e^{i\theta})| = 1$ almost everywhere.

Lemma 2.5. *Suppose that the sequences* $\{z_j\}_{-\infty}^{\infty}$, $\{\nu_j\}_{-\infty}^{\infty}$ *are such that* $-1 < z_j < z_{j+1} < 1$, $\nu_j \in N$, $j = 0, \pm 1, \dots$,

$$\sum_{j=-\infty}^{\infty} \nu_j(1 - |z_j|) < \infty,$$

and for

$$W(z) := \prod_{j=-\infty}^{\infty} \left(-\operatorname{sign} z_j \frac{z - z_j}{1 - z_j z} \right)^{\nu_j}$$

there exist $\alpha_j \in (z_j, z_{j+1})$, $j = 0, \pm 1, \dots$, *such that* $|W(\alpha_j)| \geq c > 0$. *Then for any* $f \in \mathcal{H}_p$, $1 \leq p \leq \infty$, *we have*

$$\frac{1}{2\pi i} \int_{|z|=1} \frac{f(z)}{W(z)}\, dz = \sum_{j=-\infty}^{\infty} \frac{(-\operatorname{sign} z_j)^{\nu_j}}{(\nu_j - 1)!} \left(\frac{f(z)(1 - z_j z)^{\nu_j}}{\omega_j(z)} \right)^{(\nu_j - 1)}_{\big|z=z_j},$$

where

$$\omega_j(z) := \prod_{m \neq j} \left(-\operatorname{sign} z_m \frac{z - z_m}{1 - z_m z} \right)^{\nu_m}.$$

Proof. Since $\mathcal{H}_p \subset \mathcal{H}_1$ for all $1 < p \leq \infty$, it suffices to consider the case $p = 1$. We shall assume without loss of generality that $z_j \neq 0$, $j = 0, \pm 1, \dots$ (otherwise we can achieve this situation by a conformal mapping of the unit disk). From the Fejér–Riesz (see Duren [1970, . 46])

$$\int_{-1}^{1} |f(x)|\, dx \leq \pi \|f\|_{\mathcal{H}_1}$$

and the fact that for all $x \in (0, 1)$

$$\left| \frac{ix - z_j}{1 - z_j ix} \right| \geq |z_j|,$$

and hence $|W(ix)| \geq |W(0)|$, it follows that the integrals

$$I_s := \frac{1}{2\pi i} \int_{\partial D_s} \frac{f(z)}{W(z)}\, dz, \quad s = 1, 2,$$

exist; here $D_s := \{z \in D : (-1)^s \operatorname{Re} z > 0\}$. Thus,

$$\frac{1}{2\pi i} \int_{|z|=1} \frac{f(z)}{W(z)}\, dz = I_0 + I_1.$$

Assume for definiteness that $z_0 < 0 < z_1$. We prove that

$$I_0 = \sum_{j=1}^{\infty} \frac{(-1)^{\nu_j}}{(\nu_j - 1)!} \left(\frac{f(z)(1 - z_j z)^{\nu_j}}{\omega_j(z)} \right)^{(\nu_j - 1)}_{\Big| z = z_j} . \tag{2.11}$$

Fix an arbitrary $\varepsilon > 0$ and choose $\varphi \in (0, \pi/2)$ so that

$$\frac{1}{2\pi} \int_{-\varphi}^{\varphi} |f(e^{i\theta})| \, d\theta < \varepsilon.$$

Then in view of the well-known equality (see, for example, Goluzin [1966])

$$\lim_{r \to 1} \frac{1}{2\pi} \int_{0}^{2\pi} |f(re^{i\theta}) - f(e^{i\theta})| \, d\theta = 0,$$

there exists $r_1 \in (0, 1)$ such that for all $r \in (r_1, 1)$

$$\frac{1}{2\pi} \int_{0}^{2\pi} |f(re^{i\theta}) - f(e^{i\theta})| \, d\theta < \varepsilon.$$

Thus for all $r \in (r_1, 1)$ we have

$$\begin{aligned} &\frac{1}{2\pi} \int_{-\varphi}^{\varphi} |f(re^{i\theta})| \, d\theta \\ &\qquad \le \frac{1}{2\pi} \int_{-\varphi}^{\varphi} |f(e^{i\theta})| \, d\theta + \frac{1}{2\pi} \int_{0}^{2\pi} |f(re^{i\theta}) - f(e^{i\theta})| \, d\theta < 2\varepsilon. \end{aligned} \tag{2.12}$$

From the Fejér–Riesz inequality it follows that there exists $r_2 \in (0, 1)$ such that

$$\frac{1}{2\pi} \int_{r_2}^{1} |f(xe^{i\theta})| \, dx < \varepsilon \tag{2.13}$$

for $\theta = \pm\varphi$. Let

$$r_0 := \max\{\, r_1, r_2, (1 - \sin\varphi)/(1 + \sin\varphi) \,\}, \qquad N := \max\{\, n : \alpha_n \le r_0 \,\}.$$

Then, upon setting

$$\begin{aligned} S_n &:= \{\, z \in D : |z| > \alpha_n,\ |\arg z| < \varphi \,\}, \qquad & \Omega_n &:= D_0 \setminus S_n, \\ \Gamma_n &:= \partial D \cap \partial S_n, & \gamma_n &:= D \cap \partial S_n, \end{aligned}$$

we find that for all $n > N$

$$R_n := \left| I_0 - \sum_{j=1}^{n} \frac{(-1)^{\nu_j}}{(\nu_j - 1)!} \left(\frac{f(z)(1 - z_j z)^{\nu_j}}{\omega_j(z)} \right)^{(\nu_j - 1)}_{\Big| z = z_j} \right|$$
$$= \left| I_0 - \frac{1}{2\pi i} \int_{\partial\Omega_n} \frac{f(z)}{W(z)}\, dz \right| = \left| \frac{1}{2\pi i} \int_{\Gamma_n} \frac{f(z)}{W(z)}\, dz - \frac{1}{2\pi i} \int_{\gamma_n} \frac{f(z)}{W(z)}\, dz \right|$$
$$\le \frac{1}{2\pi} \int_{\Gamma_n} |f(z)|\, |dz| + \frac{1}{2\pi} \int_{\gamma_n} \left| \frac{f(z)}{W(z)} \right| |dz| < \varepsilon + \frac{b_n}{2\pi} \int_{\gamma_n} |f(z)|\, |dz|,$$

where

$$b_n := \sup_{z \in \gamma_n} |W(z)|^{-1}.$$

It is not difficult to see that the set of points z for which

$$\left| \frac{z - z_j}{1 - z_j z} \right| < \left| \frac{\alpha_n - z_j}{1 - z_j \alpha_n} \right|,$$

for all $j \ge n + 1$ is a disk lying inside S_n, while for all $1 \le j \le n$ it is a disk lying outside S_n. Hence it follows that for all $z \in \gamma_n$ and $j \ge 0$

$$\left| \frac{z - z_j}{1 - z_j z} \right| \ge \left| \frac{\alpha_n - z_j}{1 - z_j \alpha_n} \right|.$$

Consequently, for any $z \in \gamma_n$ we have

$$|W(z)| \ge \prod_{j=0}^{-\infty} \left| \frac{z - z_j}{1 - z_j z} \right| \prod_{j=1}^{\infty} \left| \frac{\alpha_n - z_j}{1 - z_j \alpha_n} \right| \ge |W(\alpha_n)| \prod_{j=0}^{-\infty} |z_j| \ge c \prod_{j=0}^{-\infty} |z_j| =: c_1.$$

Thus, $b_n \le c_1^{-1}$ and in view of (2.12), (2.13)

$$R_n < \varepsilon + \frac{1}{2\pi c_1} \left(\int_{-\varphi}^{\varphi} |f(a_n e^{i\theta})|\, d\theta + \int_{r_0}^{1} |f(xe^{i\varphi})|\, dx + \int_{r_0}^{1} |f(xe^{-i\varphi})|\, dx \right)$$
$$< \varepsilon(1 + 4c_1^{-1}).$$

Thus, (2.11) is proved. The equality

$$I_1 = \sum_{j=0}^{-\infty} \frac{1}{(\nu_j - 1)!} \left(\frac{f(z)(1 - z_j z)^{\nu_j}}{\omega_j(z)} \right)^{(\nu_j - 1)}_{\Big| z = z_j}$$

is proved in a similar fashion. □

The analogue of Theorem 2.1 for the information operator (2.9) is the following result.

Theorem 2.6. *Suppose that the system $\{z_j\}_{-\infty}^{\infty}$ satisfies the hypothesis of Lemma* 2.5. *Then for all* $1 \le p \le \infty$ *the method*

$$f(\xi) \approx \sum_{j=-\infty}^{\infty} \sum_{\nu=0}^{\nu_j-1} c_{j\nu}(\xi,p) f^{(\nu)}(z_j),$$

where

$$c_{j\nu}(\xi,p) = \frac{W(\xi)(1-|\xi|^2)^{\frac{p-2}{p}}}{\nu!(\nu_j-\nu-1)!} \left(\frac{(1-z_j z)^{\nu_j}}{\omega_j(z)(\xi - z)(1-\bar{\xi}z)^{\frac{p-2}{p}}} \right)^{(\nu_j-\nu-1)}\Bigg|_{z=z_j},$$

is an optimal method of recovery on the class H_p and

$$e(\xi, I, H_p) = \frac{|W(\xi)|}{(1-|\xi|^2)^{1/p}}.$$

Proof. For $\xi \in D$ and $f \in H_p$ put

$$Jf := \frac{1}{2\pi i} \int_{|z|=1} \frac{W(\xi)}{W(z)(z-\xi)} \left(\frac{1-|\xi|^2}{1-\bar{\xi}z} \right)^{\frac{p-2}{p}} f(z)\,dz. \tag{2.14}$$

Consider the function $g(z) := W(z)(1-\bar{\xi}z)^{-2/p}$. We have

$$Jf = \frac{\alpha}{2\pi} \int_0^{2\pi} \overline{g(e^{i\theta})} |g(e^{i\theta})|^{p-2} f(e^{i\theta})\,d\theta, \qquad 1 \le p < \infty,$$

where $\alpha = W(\xi)(1-|\xi|^2)^{(p-2)/p}$, and for $p = \infty$

$$Jf = \frac{\alpha}{2\pi} \int_0^{2\pi} \overline{g(e^{i\theta})} \varphi(e^{i\theta}) f(e^{i\theta})\,d\theta,$$

where $\varphi(z) = |1-\bar{\xi}z|^{-2}$. On the other hand, by applying Lemma 2.5 to (2.14), we obtain

$$f(\xi) - \sum_{j=-\infty}^{\infty} \sum_{\nu=0}^{\nu_j-1} c_{j\nu}(\xi,p) f^{(\nu)}(z_j) = Jf.$$

It remains to use Theorem 1.10. □

Corollary 2.7. *Suppose that the system* $\{z_j\}_{-\infty}^{\infty}$ *satisfies the hypothesis of Lemma* 2.5 *for* $\nu_j \equiv 1$. *Then for all* $1 \le p \le \infty$ *the method*

$$f(\xi) \approx W(\xi) \sum_{j=-\infty}^{\infty} \frac{1}{W'(z_j)(\xi - z_j)} \left(\frac{1-|\xi|^2}{1-\bar{\xi} z_j} \right)^{\frac{p-2}{p}} f(z_j)$$

is an optimal method of recovery on the class H_p.

We consider some concrete systems of nodes. Let $\lambda \in (0,1)$,

$$a_j := \tanh\left(j \frac{\pi \Lambda}{\Lambda'} \right), \quad j = 0, \pm 1, \dots , \tag{2.15}$$

and

$$B_0(z) := z \prod_{j=1}^{\infty} \frac{a_j^2 - z^2}{1 - a_j^2 z^2}, \tag{2.16}$$

where Λ and Λ' are the complete elliptic integrals of the first kind for the moduli λ and $\lambda' = \sqrt{1-\lambda^2}$, respectively. Put $h = e^{-\pi\Lambda/\Lambda'}$. Then

$$a_j = \frac{1-h^{2j}}{1+h^{2j}}.$$

By making the substitution $z = -i \tan \pi v$, we have

$$z^2 = \frac{\cos 2\pi v - 1}{\cos 2\pi v + 1}$$

and

$$B_0(z) = -i \tan \pi z \prod_{j=1}^{\infty} \frac{1 - 2h^{2j} \cos 2\pi v + h^{4j}}{1 + 2h^{2j} \cos 2\pi v + h^{4j}}.$$

Using the representation of the elliptic sine and cosine by means of theta functions and their product expansions (see Appendix), we obtain

$$B_0(z) = -i\sqrt{\lambda} \frac{\operatorname{sn}(2\Lambda' v, \lambda')}{\operatorname{cn}(2\Lambda' v, \lambda')}.$$

From the second principal transform of the first degree (see Appendix)

$$\frac{\operatorname{sn}(iu, \lambda')}{\operatorname{cn}(iu, \lambda')} = i \operatorname{sn}(u, \lambda)$$

and the fact that $v = \frac{i}{\pi}$ arctanh z we find

$$B_0(z) = \sqrt{\lambda}\,\mathrm{sn}\left(\frac{2\Lambda'}{\pi}\,\mathrm{arctanh}\, z, \lambda\right). \tag{2.17}$$

Put

$$b_j := \tanh\left((2j-1)\frac{\pi\Lambda}{2\Lambda'}\right), \quad j = 0, \pm 1, \dots . \tag{2.18}$$

Then

$$B_0(b_j) = (-1)^{j+1}\sqrt{\lambda}. \tag{2.19}$$

Consequently, the system $\{a_j\}_{-\infty}^{\infty}$ satisfies the condition of Lemma 2.5, and Corollary 2.7 is applicable to it.

Denote by $H_{\infty}^{\beta}(R)$ the class of all functions analytic in the strip S_β for which $|f(z)| \leq 1$, $z \in S_\beta$. Under the conformal map of the strip S_β onto the unit disk D, defined by the function

$$z = \tanh\left(\frac{\pi}{4\beta}w\right),$$

the system of nodes $\{j\tau\}_{-\infty}^{\infty}$, $\tau > 0$, is taken to the system $\{a_j\}_{-\infty}^{\infty}$ if λ is chosen from the condition $\Lambda'/\Lambda = 4\beta/\tau$, that is, $\lambda = \kappa(2\pi\beta/\tau)$. Using the conformal equivalence of considered recovery problems for the classes $H_{\infty}^{\beta}(R)$ and H_∞, we obtain from Corollary 2.7

Corollary 2.8. *For all $\tau > 0$ and $t \in R$ the method*

$$f(t) \approx \frac{\pi}{\Lambda'}\,\mathrm{sn}\left(\frac{\Lambda'}{2\beta}t, \lambda\right) \sum_{j=-\infty}^{\infty} (-1)^j \frac{f(j\tau)}{\sinh\left(\frac{\pi}{2\beta}(t - j\tau)\right)},$$

in which $\lambda = \kappa(2\beta/\tau)$, is an optimal method of recovery on the class $H_{\infty}^{\beta}(R)$. Moreover,

$$e(t, I, H_{\infty}^{\beta}(R)) = \sqrt{\lambda}\left|\mathrm{sn}\left(\frac{\Lambda'}{2\beta}t, \lambda\right)\right|.$$

It should be noted that for the maximum value of the intrinsic error we have

$$\max_{t \in R} e(t, I, H_{\infty}^{\beta}(R)) = \sqrt{\lambda} = 2e^{-\pi\beta/\tau} + O\left(e^{-5\pi\beta/\tau}\right).$$

It is known that entire functions can be exactly reconstructed from their values at the equidistant system of points with sufficiently small step (this reconstruction is given by Kotelnikov's formula or Cartwright's formula). In contrast to this fact functions which are analytic and bounded in a strip can not be exactly recovered from their values at the equidistant system of points independently of a step. Nevertheless, Corollary 2.8 gives the possibility to recover approximately analytic functions (moreover, in the optimal way) which are given at the equidistant system of points with arbitrary step.

2.4 Optimal Nodes

In Sections 2.1 and 2.2 we obtained optimal methods of recovery of functions from the classes H_p and $H_p^\beta(T)$ at a fixed point from their values at some system of nodes. If we wish to recover a function not at one point but on some set, then it is natural to ask: how to choose a system of nodes to make the error of optimal recovery less?

Let us formulate this problem more precisely. Let W be a class of functions defined on the set G, X be a normed linear space and $W \subset X$. Put

$$I_\tau f := (f(t_1), \dots, f(t_n)), \quad \tau = (t_1, \dots, t_n) \in G^n.$$

By the problem of *optimal recovery on the space* X we mean the problem of finding the value

$$s_n(W, X) := \inf_{\tau \in G^n} \inf_{S\colon M^n \to X} \sup_{f \in W} \|f - S(I_\tau f)\|_X, \tag{2.20}$$

where $M = R$ or C (depending on the field over which the space X is considered). If some of s points coincide, we assume that together with values of function values of consecutive derivatives up to $(s-1)$-st order are given (we assume that functions from W are sufficiently smooth). Any nodes for which the lower bound in (2.20) is attained we call *optimal nodes*. A method S for which the lower bound is attained for optimal nodes we call an *optimal method of recovery on the space* X.

Proposition 2.9. *If W is a balanced class of functions, then*

$$s_n(W, X) \ge \inf_{\tau \in G^n} \sup_{\substack{f \in W \\ I_\tau f = 0}} \|f\|_X. \tag{2.21}$$

Proof. Let $\tau = (t_1, \dots, t_n) \in G^n$. For any $\varepsilon > 0$ there exists $f_\varepsilon \in W$ such that $I_\tau f_\varepsilon = 0$ and

$$\|f_\varepsilon\|_X \geq \sup_{\substack{f \in W \\ I_\tau f = 0}} \|f\|_X - \varepsilon.$$

For any method S we have

$$\|f_\varepsilon - S(0)\|_X + \| - f_\varepsilon - S(0)\|_X \geq 2\|f_\varepsilon\|_X.$$

Since W is balanced,

$$\sup_{f \in W} \|f - S(I_\tau f)\|_X \geq \|f_\varepsilon\|_X \geq \sup_{\substack{f \in W \\ I_\tau f = 0}} \|f\|_X - \varepsilon.$$

In view of arbitrariness of $\varepsilon > 0$, S, and τ we get (2.21). □

Denote by $e(t, I_\tau, W)$ the intrinsic error of optimal recovery of functions from the class W at a point $t \in G$ from the values of the information operator I_τ. In view of duality (1.11) for the convex and balanced class W

$$e(t, I_\tau, W) = \sup_{\substack{f \in W \\ I_\tau f = 0}} |f(t)|. \tag{2.22}$$

Recall that any function for which the upper bound in (2.22) is attained we call an extremal function.

Proposition 2.10. *Let W be a convex balanced class of functions. Suppose that $\varphi(t, I_\tau)$ is an optimal method of recovery of functions from this class at a point $t \in G$ using the information operator I_τ. If for all $\tau \in G^n$, $\varphi(\cdot, I_\tau) \in X$ and there exists an extremal function in the problem (2.22) which does not depend on t, then*

$$s_n(W, X) = \inf_{\tau \in G^n} \|e(\cdot, I_\tau, W)\|_X$$

and $\varphi(t, I_\tau)$ is an optimal method of recovery on the space X.

Proof. Putting $S(t, I_\tau) = \varphi(t, I_\tau)$, we have

$$s_n(W, X) \leq \inf_{\tau \in G^n} \|e(f(\cdot), W, I_\tau)\|_X.$$

For $\tau \in G^n$ denote by f_τ an extremal function for the problem (2.22) which does not depend on t. Taking into account (2.21), we obtain

$$s_n(W, X) \geq \inf_{\tau \in G^n} \|f_\tau\|_X = \inf_{\tau \in G^n} \|e(f(\cdot), W, I_\tau)\|_X.$$

□

Let E be a compact from the unit disk D and μ be a positive measure on it. Denote by $\mathcal{B}_n$ the set of all Blaschke products of degree at least n. Since the conditions of Proposition 2.10 for the class H_∞ (Theorem 2.1) hold, for all $1 \leq q \leq \infty$ we have

$$s_n(H_\infty, L_q(E,\mu)) = \inf_{B \in \mathcal{B}_n} \|B\|_{L_q(E,\mu)}. \tag{2.23}$$

In the analogous problem for the class $H_\infty^\beta(T)$ (Theorem 2.3) an extremal function does not depend on the point in which a function is recovered only for even n. Thus we have

$$s_{2n}(H_\infty^\beta(T), L_q(T)) = \inf_{\tau \in T^{2n}} \left\| k^n \prod_{j=1}^{2n} \operatorname{sn}\left(\frac{K}{\pi}(\cdot - t_j)\right)\right\|_q, \tag{2.24}$$

where $\|\cdot\|_q$ is the standard norm in $L_q(T)$, $k = \kappa(\beta)$, and K is the complete elliptic integral of the first kind with the modulus k.

We turn our attention to the case when s_n can be expressed in terms of the intrinsic error at a point in which we recover functions. In view of (2.21)

$$s_n(H_p^\beta(T), L_\infty(T)) \geq \inf_{\tau \in T^n} \sup_{f \in H_p^\beta(T)} \sup_{t \in T} |f(t)| = \inf_{\tau \in T^n} \sup_{t \in T} e(t, I_\tau, H_p^\beta(T)).$$

On the other hand, if we consider the method of recovery obtained in Theorem 2.3, we obtain

$$s_n(H_p^\beta(T), L_\infty(T)) \leq \inf_{\tau \in T^n} \sup_{t \in T} e(t, I_\tau, H_p^\beta(T)).$$

Hence for all $n \in N$ and for all $1 \leq p \leq \infty$

$$s_n(H_p^\beta(T), L_\infty(T)) = \left(\frac{2K}{\pi}\right)^{1/p} k^{[\frac{n+1}{2}]} \inf_{\tau \in T^n} \sup_{t \in T} \left| \prod_{j=1}^{n} \operatorname{sn}\left(\frac{K}{\pi}(t - t_j)\right)\right|, \tag{2.25}$$

where $[x]$ is the integral part of x.

First we consider the extremal problem (2.23) for the case when $E = T_\rho := \{z \in C : |z| = \rho\}$, $0 < \rho < 1$, and $\mu = \sigma_\rho$, where $d\sigma_\rho(\rho e^{i\theta}) = \frac{1}{2\pi} d\theta$.

Proposition 2.11. *For all* $1 \leq q \leq \infty$ *and* $0 < \rho < 1$

$$\inf_{B \in \mathcal{B}_n} \|B\|_{L_q(T_\rho, \sigma_\rho)} = \rho^n.$$

Proof. Putting $\widetilde{B}(z) = z^n$, we obtain

$$\inf_{B \in \mathcal{B}_n} \|B\|_{L_q(T_\rho,\sigma_\rho)} \le \|\widetilde{B}\|_{L_q(T_\rho,\sigma_\rho)} = \rho^n.$$

Since $\|\cdot\|_{L_1(T_\rho,\sigma_\rho)} \le \|\cdot\|_{L_q(T_\rho,\sigma_\rho)}$ for all $1 < q \le \infty$, it suffices to prove the lower bound for $q = 1$. Let $0 \le m \le n$,

$$B(z) = z^m \prod_{j=1}^{n-m} \frac{z - z_j}{1 - \overline{z}_j z},$$

and $|z_j| \le \rho$, $j = 1, \dots, s$, $|z_j| > \rho$, $j = s+1, \dots, n-m$. In view of Jensen's inequality

$$\int_{T_\rho} |B(z)|\, d\sigma_\rho(z) \ge \exp\left(\int_{T_\rho} \log|B(z)|\, d\sigma_\rho(z) \right).$$

By Jensen's formula we have

$$\begin{aligned}\int_{T_\rho} \log|B(z)|\, d\sigma_\rho(z) &= m \log\rho + \sum_{j=1}^{n-m} \log|z_j| + \sum_{j=1}^{s} \log\frac{\rho}{|z_j|} \\ &= (m+s)\log\rho + \sum_{j=s+1}^{n-m} \log|z_j| \ge n \log\rho.\end{aligned}$$

Thus for all $B \in \mathcal{B}_n$

$$\|B\|_{L_1(T_\rho,\sigma_\rho)} \ge \rho^n.$$

□

Theorem 2.12. *For all* $1 \le q \le p \le \infty$

$$s_n(H_p, L_q(T_\rho, \sigma_\rho)) = \rho^n.$$

Proof. By Propositions 2.9 and 2.11

$$s_n(H_p, L_q(T_\rho,\sigma_\rho)) \ge \inf_{\tau \in E^n} \sup_{\substack{f \in H_p \\ I_\tau f = 0}} \|f\|_{L_q(T_\rho,\sigma_\rho)} \ge \inf_{B \in \mathcal{B}_n} \|B\|_{L_q(T_\rho,\sigma_\rho)} = \rho^n.$$

Since $H_p \subset H_q$ for all $1 \le q < p \le \infty$,

$$s_n(H_p, L_q(T_\rho,\sigma_\rho)) \le s_n(H_q, L_q(T_\rho,\sigma_\rho)).$$

Therefore it suffices to prove the upper bound for the case when $p = q$. Put

$$p_f(\xi) := \sum_{j=0}^{n-1} \frac{\xi^j}{j!}(1 - |\xi|^{2(n-j)}) f^{(j)}(0).$$

Recall that $p_f(\xi)$ is an optimal method of recovery at a point ξ of function from the class H_∞ (see (2.3)). By the residue theorem we have

$$f(\xi) - p_f(\xi) = \frac{1}{2\pi i} \int_{|z|=1} \frac{\xi^n (1 - |\xi|^2) f(z)}{z^n (z - \xi)(1 - \bar{\xi} z)} \, dz$$
$$= \frac{1}{2\pi} \int_0^{2\pi} \frac{\xi^n (1 - |\xi|^2) f(e^{i\theta})}{e^{in\theta} |1 - e^{-i\theta}\xi|^2} \, d\theta.$$

Thus for $\xi = \rho e^{it}$ we get

$$f(\xi) - p_f(\xi) = \rho^n (P_\rho * F)(t),$$

where

$$P_\rho(\theta) = \frac{1}{2\pi} e^{in\theta} \frac{1 - \rho^2}{|1 - \rho e^{i\theta}|^2}, \qquad F(\theta) = f(e^{i\theta}).$$

Using the well-known inequality

$$\|f * g\|_q \le \|f\|_1 \|g\|_q,$$

we have

$$s_n(H_q, L_q(T_\rho, \sigma_\rho)) \le \sup_{f \in H_q} \|f - p_f\|_{L_q(T_\rho, \sigma_\rho)} \le \rho^n \|P_\rho\|_1 = \rho^n.$$

(Since $|P_\rho(\theta)| = \dfrac{1}{2\pi} P(e^{i\theta}, \rho)$, where P is the Poisson kernel, $\|P_\rho\|_1 = 1$). □

In order to solve the extremal problem (2.23) in the case when E is an interval of real axis and extremal problems (2.24), (2.25), we need a generalization of a result of Pinkus [1979].

We denote by $S(f)$ the number of sign changes of a picewise-continuous real 2π-periodic function over a period. For a real continuous 2π-periodic function $\mathcal{K}$ we set

$$(\mathcal{K} * f)(x) := \int_0^{2\pi} \mathcal{K}(x - t) f(t) \, dt.$$

We say that $\mathcal{K} \in NCVD$ (non-degenerate cyclic variation diminishing), if $S(\mathcal{K} * f) \le S(f)$ for all f and

$$\dim \operatorname{span}\{\, \mathcal{K}(x_1 - \cdot), \dots, \mathcal{K}(x_n - \cdot)\,\} = n$$

for all $0 \le x_1 < \dots < x_n < 2\pi$ and all n. If for all $0 \le x_1 < \dots < x_{2l+1} < 2\pi$, $0 \le y_1 < \dots < y_{2l+1} < 2\pi$ the inequality

$$\varepsilon \det \left(\mathcal{K}(x_j - y_m)\right)_{j,m=1}^{2l+1} > 0$$

holds, where $\varepsilon = 1$ or -1, then we say that $\mathcal{K} \in SSC_{2l+1}$ (strictly sign consistent).

Put

$$\Theta_n := \{\, \theta : \theta = (\theta_1, \dots, \theta_n),\ 0 \le \theta_1 \le \dots \le \theta_n < 2\pi \,\}.$$

For $\eta \in \Theta_{2n}$ we set

$$h_\eta(t) := (-1)^j, \quad t \in [\eta_{j-1}, \eta_j), \quad j = 1, \dots, 2n+1,$$

where $\eta_0 := 0$, $\eta_{2n+1} := 2\pi$. We denote by $h_n(t)$ the function $h_\eta(t)$, when $\eta_j = (j-1)\pi/n$, $j = 1, \dots, 2n$.

Theorem 2.13. *Let $\mathcal{K} \in NCVD$, and let φ be a nonnegative continuously differentiable function defined on $[0, \|\mathcal{K}\|_1]$ such that φ' is positive and increasing in the interval $(0, \|\mathcal{K}\|_1)$. Then for all $1 \le q \le \infty$*

$$\inf_{\eta \in \Theta_{2n}} \|\varphi(|\mathcal{K} * h_\eta|)\|_q = \|\varphi(|\mathcal{K} * h_n|)\|_q.$$

Moreover, if $\mathcal{K} \in SSC_{2l+1}$, $l = 0, 1, \dots, n$, and the infimum is attained for $\eta^* \in \Theta_{2n}$, then $\eta^*_{j+1} - \eta^*_j = \pi/n$, $j = 1, \dots, 2n-1$.

This theorem was proved for $\varphi(x) = x$ by Pinkus [1979] (see also Pinkus [1985, p. 174]). The proof for the general case is carried out along the same lines.

Theorem 2.14. *Let φ be a function defined on $[0,1]$ and satisfying the condition of Theorem 2.13. Then for all $k \in (0,1)$ and $s \in N$*

$$\inf_{\theta \in \Theta_s} \left\| \varphi\left(\frac{4}{\pi} \arctan \left| k^{s/2} \prod_{j=1}^{s} \operatorname{sn}\left(\frac{K}{\pi}(\cdot - \theta_j)\right)\right|\right)\right\|_q$$

$$= \begin{cases} \left(\dfrac{2\pi}{\Lambda} \displaystyle\int_0^1 \frac{\varphi^q\left(\frac{4}{\pi}\arctan(\sqrt{\lambda} t)\right) dt}{\sqrt{(1-t^2)(1-\lambda^2 t^2)}} \right)^{1/q}, & 1 \le q < \infty, \\ \varphi\left(\dfrac{4}{\pi}\arctan\sqrt{\lambda}\right), & q = \infty, \end{cases}$$

where $\lambda = \kappa\left(\frac{\pi K'}{2K}s\right)$, κ *is defined by* (2.5), *and* Λ *is the complete elliptic integral of the first kind for the modulus* λ *(that is,* λ *is defined by the condition* $\Lambda'/\Lambda = sK'/K$*). Moreover, if the infimum is attained for* θ^**, then* $\theta^*_{j+1} - \theta^*_j = 2\pi/s$, $j = 1, \dots, s-1$.

Proof. We denote by $h^{\beta}_{\infty,R}(T)$ the class of analytic functions in the strip S_β that are real on the real axis, 2π-periodic and satisfy the condition

$$|\operatorname{Re} f(z)| \le 1, \quad z \in S_\beta.$$

Functions from the class $h^{\beta}_{\infty,R}(T)$ admit the representation (see Akhiezer [1965, p. 269])

$$f(z) = \int_0^{2\pi} \mathcal{K}_\beta(z-t) \operatorname{Re} f(t+i\beta)\, dt,$$

where

$$\mathcal{K}_\beta(z) = \frac{1}{2\pi}\left(1 + 2\sum_{j=1}^{\infty} \frac{\cos jz}{\cosh j\beta}\right).$$

It is known (see Pinkus [1985, p. 128]), that $\mathcal{K}_\beta \in NCVD$. Furthermore, it was proved by Forst [1977] that $\mathcal{K}_\beta \in SSC_{2l+1}$ for all $l = 0, 1, \dots$.

First we consider the case when s is even. Let $s = 2n$ and $\theta \in \Theta_{2n}$. Put

$$f_\theta(z) := \frac{4}{\pi}\arctan\left(k^n \prod_{j=1}^{2n} \operatorname{sn}\left(\frac{K}{\pi}(z-\theta_j)\right)\right).$$

Since $\operatorname{sn}(u+2K) = -\operatorname{sn} u$ and $|\sqrt{k}\operatorname{sn} u| < 1$ for $|\operatorname{Im} u| < K'/2$, it follows that $f_\theta \in h^{\beta}_{\infty,R}(T)$ for $\beta = \dfrac{\pi K'}{2K}$. It follows from the equality

$$\sqrt{k}\operatorname{sn}(u + iK'/2) = \frac{(1+k)\operatorname{sn} u + i \operatorname{cn} u \operatorname{dn} u}{1 + k\operatorname{sn}^2 u} \tag{2.26}$$

(see Appendix) that

$$\sqrt{k}\operatorname{sn}(u + iK'/2) = e^{i\omega(u)}$$

for $0 \le u \le 2K$, where $\omega(u)$ is monotone increasing from 0 to π. Since for $|z| = 1$ and $z \ne \pm i$

$$\operatorname{Re}\left(\frac{4}{\pi}\arctan z\right) = \operatorname{sign} \operatorname{Re} z,$$

it follows that for all $t \in T$

$$\operatorname{Re} f_\theta(t+i\beta) = \operatorname{sign} \operatorname{Re} \exp\left(i\sum_{j=1}^{2n} \omega_j(t)\right),$$

where $\sum_{j=1}^{2n} \omega_j(t)$ is monotone increasing from some α to $\alpha + 2\pi n$. Hence there exists $\eta \in \Theta_{2n}$ such that

$$f_\theta(z) = \varepsilon(\mathcal{K}_\beta * h_\eta)(z)$$

for $\varepsilon = 1$ or -1. From Theorem 2.13 for $\mathcal{K} = \mathcal{K}_\beta$ we obtain

$$\inf_{\theta\in\Theta_{2n}} \|\varphi(|f_\theta(\cdot)|)\|_q \geq \inf_{\eta\in\Theta_{2n}} \|\varphi(|(\mathcal{K}_\beta * h_\eta)(\cdot)|)\|_q = \|\varphi(|(\mathcal{K}_\beta * h_n)(\cdot)|)\|_q. \tag{2.27}$$

We denote by f_n the function f_θ for $\theta_j = (j-1)\pi/n$, $j = 1, \dots, 2n$. Using the equalities

$$\operatorname{sn}(u - 2K) = -\operatorname{sn} u, \qquad \operatorname{sn}(u+v)\operatorname{sn}(u-v) = \frac{\operatorname{sn}^2 u - \operatorname{sn}^2 v}{1 - k^2 \operatorname{sn}^2 u \operatorname{sn}^2 v},$$

as well as the first principal transform of elliptic functions of degree $2n$ (see Appendix) we find

$$\begin{aligned}
&f_n\left(z - \frac{\pi}{2n}\right) \\
&= \frac{4}{\pi}\arctan\left((-1)^n k^n \prod_{j=1}^{n} \operatorname{sn}\left(\frac{K}{\pi}z - \frac{2j-1}{2n}K\right)\operatorname{sn}\left(\frac{K}{\pi}z + \frac{2j-1}{2n}K\right)\right) \\
&= \frac{4}{\pi}\arctan\left(k^n \prod_{j=1}^{n} \frac{\operatorname{sn}^2 \frac{2j-1}{2n}K - \operatorname{sn}^2 \frac{K}{\pi}z}{1 - k^2 \operatorname{sn}^2 \frac{2j-1}{2n}K \operatorname{sn}^2 \frac{K}{\pi}z}\right) \\
&= \frac{4}{\pi}\arctan\left(\sqrt{\lambda}\operatorname{sn}\left(\frac{2n\Lambda}{\pi}z + \Lambda, \lambda\right)\right),
\end{aligned}$$

where $\Lambda'/\Lambda = 2nK'/K$, that is, $\lambda = \kappa\left(\frac{\pi K'}{K}n\right)$ (see (A.38)). Thus,

$$f_n(z) = -\frac{4}{\pi}\arctan\left(\sqrt{\lambda}\operatorname{sn}\left(\frac{2n\Lambda}{\pi}z, \lambda\right)\right).$$

Since $f_n \in h^{\beta}_{\infty,R}(T)$ and in view of (2.26),

$$\operatorname{Re} f_n(t+i\beta) = -\operatorname{sign}\operatorname{sn}\left(\frac{2n\Lambda}{\pi}z,\lambda\right) = -h_n(t),$$

it follows that

$$(\mathcal{K}_\beta * h_n)(z) = \frac{4}{\pi}\arctan\left(\sqrt{\lambda}\operatorname{sn}\left(\frac{2n\Lambda}{\pi}z,\lambda\right)\right), \tag{2.28}$$

i.e., $f_n(z) = -(\mathcal{K}_\beta * h_n)(z)$. Thus, taking (2.27) into account, we have

$$\inf_{\theta\in\Theta_{2n}} \|\varphi(|f_\theta(\cdot)|)\|_q = \|\varphi(|f_n(\cdot)|)\|_q = \left\|\varphi\left(\frac{4}{\pi}\arctan\left|\sqrt{\lambda}\operatorname{sn}\left(\frac{2n\Lambda}{\pi}\cdot,\lambda\right)\right|\right)\right\|_q.$$

From this we immediately obtain the assertion in the case under consideration when $q=\infty$. For $1\le q<\infty$ we make the substitutions $x = \dfrac{2n\Lambda}{\pi}z$ and $t = \operatorname{sn}(x,\lambda)$, and use the standard properties of the elliptic functions to obtain

$$\begin{aligned}
&\left\|\varphi\left(\frac{4}{\pi}\arctan\left|\sqrt{\lambda}\operatorname{sn}\left(\frac{2n\Lambda}{\pi}\cdot,\lambda\right)\right|\right)\right\|_q \\
&\qquad = \left(\frac{2\pi}{\Lambda}\int_0^1 \varphi^q\left(\frac{4}{\pi}\arctan\left(\sqrt{\lambda}\operatorname{sn}(x,\lambda)\right)\right)dx\right)^{1/q} \\
&\qquad = \left(\frac{2\pi}{\Lambda}\int_0^1 \frac{\varphi^q\left(\frac{4}{\pi}\arctan(\sqrt{\lambda}t)\right)dt}{\sqrt{(1-t^2)(1-\lambda^2t^2)}}\right)^{1/q}.
\end{aligned}$$

If the infimum on the left-hand side of (2.27) is attained for $\theta\in\Theta_{2n}$, then it follows from Theorem 2.13 that f_θ is the same as $\mathcal{K}_\beta * h_n$ to within a translation. Hence the zeros of f_θ are uniformly distributed with step size π/n.

Suppose now that s is odd. Using the Landen transform (see Appendix), we find that

$$\sqrt{k}\operatorname{sn}\left(\frac{2K}{\pi}u\right) = -l\operatorname{sn}\left(\frac{L}{\pi}u,l\right)\operatorname{sn}\left(\frac{L}{\pi}(u-\pi),l\right), \tag{2.29}$$

where

$$l = \frac{2\sqrt{k}}{1+k}, \qquad \frac{L'}{L} = \frac{K'}{2K},$$

and L and L' are the complete elliptic integrals of the first kind for the moduli l and $l' = \sqrt{1-l^2}$, respectively. It follows from (2.29) that

$$k^{s/2}\prod_{j=1}^{s}\operatorname{sn}\left(\frac{K}{\pi}(t-\theta_j)\right) = (-1)^s l^s \prod_{j=1}^{2s}\operatorname{sn}\left(\frac{L}{\pi}\left(\frac{t}{2}-\frac{\theta_j}{2}\right),l\right),$$

where $\theta_{n+j} = 2\pi + \theta_j$, $j = 1,\dots,n$. The assertion of the theorem now follows from the corresponding assertion for the even case. □

Let $L_{qk} := L_q([-\sqrt{k},\sqrt{k}],\nu_k)$, where

$$d\nu_k(z) := \frac{\pi}{2K}\frac{dz}{\sqrt{(k-z^2)(1-kz^2)}}, \qquad k\in(0,1).$$

Using the theorem proved above, we can find the Blaschke products from $\mathcal{B}_n$ which have the least deviation from zero in the norm of the space L_{qk}.

Theorem 2.15. *Let φ be a function defined on $[0,1]$ and satisfying the condition of Theorem 2.13. Then for all $k\in(0,1)$*

$$\inf_{B\in\mathcal{B}_n}\left\|\varphi\left(\frac{4}{\pi}\arctan|B(\cdot)|\right)\right\|_{L_{qk}} = \begin{cases}\left(\dfrac{\pi}{\Lambda}\displaystyle\int_0^1\frac{\varphi^q\left(\frac{4}{\pi}\arctan(\sqrt{\lambda}t)\right)dt}{\sqrt{(1-t^2)(1-\lambda^2t^2)}}\right)^{1/q}, & 1\le q<\infty,\\ \varphi\left(\dfrac{4}{\pi}\arctan\sqrt{\lambda}\right), & q=\infty,\end{cases} \tag{2.30}$$

where $\lambda = \kappa\left(\frac{\pi K'}{2K}n\right)$. The unique function from $\mathcal{B}_n$ for which the infimum is attained is the Blaschke product with the zeros

$$z_j := \sqrt{k}\operatorname{sn}\left(\left(\frac{2j-1}{n}-1\right)K\right), \quad j=1,\dots,n. \tag{2.31}$$

Proof. If $x\in(-1,1)$, $\alpha+i\beta\in D$, $\alpha,\beta\in R$, and $\beta\ne0$, then

$$\left|\frac{x-\alpha-i\beta}{1-(\alpha-i\beta)x}\right| > \left|\frac{x-\alpha}{1-\alpha x}\right|. \tag{2.32}$$

Therefore the infimum in (2.30) can be considered over the Blaschke products with real zeros. Moreover, since as $\sqrt{k} < |\alpha| < 1$, for all $x \in [-\sqrt{k}, \sqrt{k}]$

$$\left|\frac{x-\alpha}{1-\alpha x}\right| > \left|\frac{x-\sqrt{k}\,\mathrm{sign}\,\alpha}{1-x\sqrt{k}\,\mathrm{sign}\,\alpha}\right|,$$

it follows that the infimum in (2.30) can be considered only over the Blaschke products with zeros from $[-\sqrt{k}, \sqrt{k}]$. Let

$$B(z) = \prod_{j=1}^{n} \frac{z-\alpha_j}{1-\alpha_j z}, \quad \alpha_j \in [-\sqrt{k}, \sqrt{k}].$$

Put $l = 2\sqrt{k}/(1+k)$. Making the substitutions $z = (x-\sqrt{k})/(1-\sqrt{k}x)$, $x = l\,\mathrm{sn}^2(y,l)$ and using the fact that $L = (1+k)K$, where L is the complete elliptic integral of the first kind for the modulus l (see the Gauss transform in Appendix), we obtain $d\nu_k(z) = \dfrac{\pi}{2L}\,dy$,

$$B(z) = l^n \prod_{j=1}^{n} \frac{\mathrm{sn}^2(y,l) - \mathrm{sn}^2(y_j,l)}{1 - l^2\,\mathrm{sn}^2(y_j,l)\,\mathrm{sn}^2(y,l)} = l^n \prod_{j=1}^{n} \mathrm{sn}(y-y_j,l)\,\mathrm{sn}(y+y_j,l),$$

where $\alpha_j = (x_j - \sqrt{k})/(1-\sqrt{k}x_j)$ and $x_j = l\,\mathrm{sn}^2(y_j,l)$. Putting $t = \pi y/L$, $\theta_j = \pi y_j/L$, we have

$$\inf_{B\in\mathcal{B}_n} \left\|\varphi\left(\frac{4}{\pi}\arctan|B(\cdot)|\right)\right\|_{L_{qk}} = \inf_{0\le\theta_j\le\pi} \left(\int_0^{\pi} \varphi^q\left(\frac{4}{\pi}\arctan|f_\theta(t)|\right)dt\right)^{1/q},$$

where

$$f_\theta(t) = l^n \prod_{j=1}^{n} \mathrm{sn}\left(\frac{L}{\pi}(t-\theta_j), l\right)\mathrm{sn}\left(\frac{L}{\pi}(t+\theta_j), l\right).$$

Since $f_\theta(-t) = f_\theta(t)$,

$$\inf_{B\in\mathcal{B}_n} \left\|\varphi\left(\frac{4}{\pi}\arctan|B(\cdot)|\right)\right\|_{L_{qk}} = 2^{-1/q} \inf_{0\le\theta_j\le\pi} \left\|\varphi\left(\frac{4}{\pi}\arctan|f_\theta(\cdot)|\right)\right\|_q. \tag{2.33}$$

Taking into account that

$$\frac{L'}{L} = \frac{K'}{2K},$$

we obtain from Theorem 2.14 the equality (2.30). Moreover, the system of nodes

$$\theta_j = \frac{2j-1}{2n}\pi, \quad j = 1, \dots, n,$$

are the unique system for which the infimum in the right-hand side of (2.33) is attained. Consequently, the unique function from $\mathcal{B}_n$ for which the infimum is attained is the Blaschke product with the zeros $z_j = (x_j - \sqrt{k})/(1-\sqrt{k}x_j)$, where

$$x_j = l \operatorname{sn}^2\left(\frac{2j-1}{2n}L, l\right).$$

Using the Gauss transform it is easy to check that the nodes z_j coincide with the nodes defined by (2.31). □

Now we can find the value $s_n(H_\infty, L_{qk})$. Putting in Theorem 2.15 $\varphi(x) = \tan\frac{\pi}{4}x$ and taking into account (2.23), we have

Theorem 2.16. *For all $k \in (0,1)$*

$$s_n(H_\infty, L_{qk}) = \begin{cases} \sqrt{\lambda}\left(\frac{\pi}{\Lambda}J_q(\lambda)\right)^{1/q}, & 1 \le q < \infty, \\ \sqrt{\lambda}, & q = \infty, \end{cases}$$

where

$$J_q(\lambda) := \int_0^1 \frac{t^q\,dt}{\sqrt{(1-t^2)(1-\lambda^2t^2)}},$$

$\lambda = \kappa\left(\frac{\pi K'}{2K}n\right)$. *Moreover, the unique optimal nodes are the nodes defined by* (2.31).

Using the expansion (see (A.36))

$$\Lambda = \frac{\pi}{2}\left(1 + 2\sum_{m=1}^{\infty} e^{-\frac{\pi\Lambda'}{\Lambda}m^2}\right)^2,$$

and the fact that

$$\int_0^1 x^q(1-x^2)^{-1/2}dx = \frac{\sqrt{\pi}}{2}\frac{\Gamma\left(\frac{q+1}{2}\right)}{\Gamma\left(\frac{q}{2}+1\right)},$$

we obtain the following asymptotic equality

$$\sqrt{\lambda}\left(\frac{\pi}{\Lambda}J_q(\lambda)\right)^{1/q} = 2\left(\sqrt{\pi}\frac{\Gamma\left(\frac{q+1}{2}\right)}{\Gamma\left(\frac{q}{2}+1\right)}\right)^{1/q} e^{-\frac{\pi K'}{4K}n} + O\left(e^{-\frac{5\pi K'}{4K}n}\right). \quad (2.34)$$

Let $\mathcal{E}_c$ be the interior of the ellipse with foci at the points ± 1 and sum $c > 1$ of its semiaxes. Denote by $H_\infty(\mathcal{E}_c)$ the set of all functions analytic in $\mathcal{E}_c$ and satisfying there the condition $|f(z)| \le 1$. The function

$$z(w) := \sqrt{k}\,\mathrm{sn}\left(\frac{2K}{\pi}\arcsin w\right), \quad (2.35)$$

where the modulus k satisfies

$$\frac{K'}{K} = \frac{4}{\pi}\log c,$$

maps $\mathcal{E}_c$ conformally on the unit disk D. This map carries the interval $[-1,1]$ to the interval $[-\sqrt{k},\sqrt{k}]$, and the *Chebyshev nodes*

$$w_j = \cos\frac{2j-1}{2n}\pi, \quad j = 1,\dots,n,$$

to the nodes (2.31). In addition,

$$d\nu_k(z) = \frac{dw}{\sqrt{1-w^2}}.$$

Therefore from Theorem 2.16 follows

Corollary 2.17. *For all* $c > 1$

$$s_n(H_\infty(\mathcal{E}_c), L_q([-1,1],\mu)) = \begin{cases} \sqrt{\lambda}\left(\frac{\pi}{\Lambda}J_q(\lambda)\right)^{1/q} = 2\left(\sqrt{\pi}\frac{\Gamma\left(\frac{q+1}{2}\right)}{\Gamma\left(\frac{q}{2}+1\right)}\right)^{1/q} c^{-n} \\ \qquad + O\left(c^{-5n}\right), & 1 \le q < \infty, \\ \sqrt{\lambda} = 2c^{-n} + O\left(c^{-5n}\right), & q = \infty, \end{cases}$$

where

$$d\mu(w) := \frac{dw}{\sqrt{1-w^2}}, \qquad \lambda = \kappa(2n\log c).$$

Moreover, the unique system of optimal nodes is the Chebyshev system of nodes.

Now let us find the value $s_n(H_p^\beta(T), L_q(T))$.

Theorem 2.18. *For all $\beta > 0$*

$$s_n(H_p^\beta(T), L_\infty(T)) = \begin{cases} \left(\frac{2K}{\pi}\right)^{1/p} \sqrt{\lambda_n}, & n \text{ even}, \\ \left(\frac{2K}{\pi}\right)^{1/p} \sqrt{k}\sqrt{\lambda_n}, & n \text{ odd}, \end{cases} \tag{2.36}$$

$$s_{2n}(H_\infty^\beta(T), L_q(T)) = \sqrt{\lambda_{2n}} \left(\frac{2\pi}{\Lambda_{2n}} J_q(\lambda_{2n})\right)^{1/q}, \quad 1 \le q < \infty, \tag{2.37}$$

where $\lambda_n = \kappa(\beta n)$, while K, K' and Λ_n are the complete elliptic integrals for the moduli $k = \kappa(\beta)$, $k' = \sqrt{1-k^2}$, and λ_n. The equidistant system of nodes on T is the unique system of optimal nodes for the value (2.36) *up to within a translation. The corresponding optimal methods of recovery have the form*

$$f(t) \approx \frac{K}{n\Lambda_n} \operatorname{sn}\left(\frac{n\Lambda_n}{\pi}t, \lambda_n\right) \sum_{j=0}^{n-1} (-1)^j c_{np} \left(\frac{K}{\pi}t - j\frac{2K}{n}\right) f\left(j\frac{2\pi}{n}\right),$$

$$f(t) \approx \frac{K}{2n\Lambda_{2n}} \operatorname{sn}\left(\frac{2n\Lambda_{2n}}{\pi}t, \lambda_{2n}\right) \sum_{j=0}^{2n-1} (-1)^j c_{2n,\infty} \left(\frac{K}{\pi}t - j\frac{K}{n}\right) f\left(j\frac{\pi}{n}\right),$$

where c_{np} are defined by (2.8).

Proof. The equalities (2.36) and (2.37) follow from (2.25), (2.24), and Theorem 2.14 in which $\varphi(x) = \tan\frac{\pi}{4}x$. To find optimal methods of recovery we use Corollary 2.4 and the fact that in view of the first principal transform of degree n (see Appendix) for the function

$$W(t) = k^{n/2} \prod_{j=0}^{n-1} \operatorname{sn}\left(\frac{K}{\pi}t - j\frac{2K}{n}\right)$$

the equality

$$W(t) = (-1)^{n+1}\sqrt{\lambda_n}\,\mathrm{sn}\left(\frac{n\Lambda_n}{\pi}t, \lambda_n\right)$$

holds. □

Using the equality analogous to (2.34), one can obtain the following asymptotic equality for $1 \le q < \infty$

$$s_{2n}(H^\beta_\infty(T), L_q(T)) = 2\left(2\sqrt{\pi}\frac{\Gamma\left(\frac{q+1}{2}\right)}{\Gamma\left(\frac{q}{2}+1\right)}\right)^{1/q} e^{-\beta n} + O\left(e^{-5\beta n}\right).$$

2.5 Functions of Several Variables

In this section, we consider the problem of optimal recovery for the multidimensional case. Let B_n be the unit ball of C^n and S_n its boundary:

$$B_n := \{\, z = (z_1, \dots, z_n) \in C^n : |z|^2 := \sum_{k=1}^{n} |z_k|^2 < 1 \,\},$$
$$S_n := \{\, z \in C^n : |z| = 1 \,\}.$$

The *Hardy space* $\mathcal{H}_p(B_n)$ is the set of holomorphic functions in B_n satisfying the condition

$$\|f\|_{\mathcal{H}_p(B_n)} := \sup_{0<r<1}\left(\int_{S_n} |f(rz)|^p \, d\sigma_n(z)\right)^{1/p} < \infty, \quad 1 \le p < \infty,$$
$$\|f\|_{\mathcal{H}_\infty(B_n)} := \sup_{z\in B_n} |f(z)| < \infty, \quad p = \infty,$$

where σ_n is the probability measure which is invariant with respect to the orthogonal group $O(2n)$. It is known (see Rudin [1980]) that functions from $\mathcal{H}_p(B_n)$ have finite boundary values almost everywhere which belong to $L_p(S_n, \sigma_n)$. Thus $\mathcal{H}_p(B_n)$ can be considered as a subspace of $L_p(S_n, \sigma_n)$.

Set

$$H_p(B_n) := \{\, f \in \mathcal{H}_p(B_n) : \|f\|_{\mathcal{H}_p(B_n)} \le 1 \,\}.$$

For multiindex $\alpha = (\alpha_1, \dots, \alpha_n)$ set

$$|\alpha| := \alpha_1 + \cdots + \alpha_n, \quad D^\alpha := D_1^{\alpha_1} \dots D_n^{\alpha_n}, \quad D_j := \frac{\partial}{\partial z_j}.$$

We consider the problem of optimal recovery of $f \in H_p(B_n)$ at a point $a \in B_n$ from exact values of the traces of $D^\alpha f$, $|\alpha| = 0, \dots, r-1$, given on some set $A \subset B$. Thus the information operator in the considered case has the form

$$If = I_A^r f := \{D^\alpha f_{|A}\}_{|\alpha|=0}^{r-1}.$$

We wish to find the intrinsic error which we denote by $e(a, I_A^r, H_p(B_n))$ and an optimal method of recovery.

We need several auxiliary results related to weighted reproducing kernels in $H_p(B_n)$. Let $u \in C$, $|u| < 1$, and $\rho \geq 1$. Set

$$\Phi_n(\rho, u) := \sum_{k=0}^{\infty} \frac{\Gamma(n+k+\rho/2)}{\Gamma(k+\rho/2+1)} u^k. \tag{2.38}$$

For $z = (z_1, \dots, z_n) \in C^n$ and $1 \leq k \leq n$ put

$$z_k' := (z_1, \dots, z_{n-k}, 0, \dots, 0), \qquad z_k'' := z - z_k',$$

$$\langle z, w \rangle := \sum_{k=1}^{n} z_k \overline{w}_k, \quad z, w \in C^n, \qquad s_k(z, w) := \frac{\langle z, w_k'' \rangle}{1 - \langle z, w_k' \rangle},$$

$$K_{rk}(z, w) := \frac{\langle z, w_k'' \rangle^r \Phi_n(rp, s_k(z, w))}{(n-1)!(1 - \langle z, w_k' \rangle)^{n+rp/2}}.$$

Note that $s_k(z, w) = \overline{s_k(w, z)}$ and $K_{rk}(z, w) = \overline{K_{rk}(w, z)}$.

Proposition 2.19. *For all $1 \leq p < \infty$, $z \in B_n$, and $f \in \mathcal{H}_p(B_n)$*

$$f(z) - \sum_{j=0}^{r-1} \frac{1}{j!} \left(D_n^j f\right)(z_1') z_n^j = \int_{S_n} K_{r1}(z, w) f(w) |w_n|^{r(p-2)} \, d\sigma_n(w). \tag{2.39}$$

Proof. As polynomials are dense in $\mathcal{H}_p(B_n)$ it is sufficient to prove (2.39) for the functions of the kind $f(z) = g(z') z_n^m$, $m = 0, 1, \dots$, where $g(z')$ is a polynomial which depends only on $z_1, \dots, z_{n-1}$ ($z' := (z_1, \dots, z_{n-1})$). Let ν_n be the normalized Lebesgue measure on B_n. The integral in (2.39) can be reduced to the integral over B_{n-1} (see Rudin [1980])

$$\int_{S_n} K_{r1}(z,w)g(w')w_n^m|w_n|^{r(p-2)}\,d\sigma_n(w)$$
$$= \frac{1}{(n-1)!}\int_{B_{n-1}} \frac{(1-|w'|^2)^{r(p-2)/2}g(w')\,d\nu_{n-1}(w')}{(1-\langle z',w'\rangle)^{n+rp/2}}$$
$$\times \frac{1}{2\pi}\int_{-\pi}^{\pi} z_n^r\overline{w}_n^r w_n^m e^{i(m-r)\theta}\sum_{k=0}^{\infty}\frac{\Gamma(n+k+rp/2)}{\Gamma(k+rp/2+1)}\frac{z_n^k\overline{w}_n^k e^{-ik\theta}}{(1-\langle z',w'\rangle)^k}\,d\theta$$
$$= \begin{cases} 0, & m<r, \\ z_n^m \displaystyle\int_{B_{n-1}} K_s(z',w')g(w')\,d\nu_{n-1}(w'), & m\ge r, \end{cases}$$

where

$$K_s(z',w') := \frac{\Gamma(n+s)}{\Gamma(n)\Gamma(s+1)}\frac{(1-|w'|^2)^s}{(1-\langle z',w'\rangle)^{n+s}}, \quad s := m + r(p-2)/2.$$

It is known that for all $s > -1$ the kernel $K_s(z',w')$ is the reproducing kernel in $H_\infty(B_{n-1})$ (see Rudin [1980, p. 121]). Thus

$$\int_{S_n} K_{r1}(z,w)g(w')w_n^m|w_n|^{r(p-2)}\,d\sigma_n(w) = \begin{cases} 0, & m<r, \\ g(z')z_n^m, & m\ge r, \end{cases}$$

It is easy to check that these values coincide with the left-hand side of (2.39). □

For $a \in C$ denote by P_a the orthogonal projection of C^n onto subspace induced by the vector a:

$$P_a := \begin{cases} \dfrac{\langle z,a\rangle}{\langle a,a\rangle}a, & a\ne 0, \\ 0, & a=0. \end{cases}$$

Proposition 2.20. *For all* $1\le p<\infty$, $z\in B_n$, *and* $f\in \mathcal{H}_p(B_n)$

$$f(z) - \sum_{j=0}^{r-1}\frac{1}{j!}d^j f_{|z_k'} = \int_{S_n} K_{rk}(z,w)f(w)|P_{z_k''}w|^{r(p-2)}\,d\sigma_n(w),$$

where $dz = z_k''$.

Proof. For $z''_k = 0$ the statement is evident. Suppose that $z''_k \neq 0$. By Proposition 2.19 it follows that for all $g \in \mathcal{H}_p(B_n)$ and $v \in B_n$

$$g(v) - \sum_{j=0}^{r-1} \frac{1}{j!} d^j g\big|_{v'_1} = \int_{S_n} K_{r1}(v,y) g(y) |y_n|^{r(p-2)} \, d\sigma_n(y), \tag{2.40}$$

where $dv = v''_1$. Let U be an unitary $n \times n$ matrix and $f \in \mathcal{H}_p(B_n)$. Set $g(v) = f(U^{-1}v)$. Then $g \in \mathcal{H}_p(B_n)$. Using (2.40), we obtain

$$f(U^{-1}v) - \sum_{j=0}^{r-1} \frac{1}{j!} d^j f\big|_{U^{-1}v'_1} = \int_{S_n} K_{r1}(v,y) f(U^{-1}y) |y_n|^{r(p-2)} \, d\sigma_n(y), \tag{2.41}$$

where $dz = U^{-1}v''_1$. For a fixed $z \in B_n$ such that $z''_k \neq 0$ consider the matrix U of the form

$$U = \begin{pmatrix} 1 & & & \\ & \ddots & & 0 \\ & & 1 & \\ & & & \\ & 0 & & C \end{pmatrix}, \tag{2.42}$$

where C is the unitary $k \times k$ matrix which transforms $(z_{n-k+1}, \cdots, z_n)$ into $(0, \ldots, 0, |z''_k|)$. Substituting $v = Uz$, $y = Uw$ in (2.41) and using the fact that the measure σ_n is invariant with respect to unitary transformations, we obtain

$$f(z) - \sum_{j=0}^{r-1} \frac{1}{j!} d^j f\big|_{z'_k} = \int_{S_n} K_{r1}(Uz, Uw) f(w) |(Uw)_n|^{r(p-2)} \, d\sigma_n(w),$$

where $dz = z''_k$. We have

$$\langle (Uz)'_1, (Uw)'_1 \rangle = \langle z'_k, (Uw)'_1 \rangle = \langle z'_k, Uw \rangle = \langle Uz'_k, Uw \rangle = \langle z'_k, w \rangle = \langle z'_k, w'_k \rangle,$$

$$\begin{aligned} |z''_k| \overline{(Uw)}_n = (Uz)_n \overline{(Uw)}_n &= \langle Uz, Uw \rangle - \langle (Uz)'_1, (Uw)'_1 \rangle \\ &= \langle z, w \rangle - \langle z'_k, w'_k \rangle = \langle z''_k, w''_k \rangle. \end{aligned}$$

Hence

$$s_1(Uz, Uw) = s_k(z, w), \quad K_{r1}(Uz, Uw) = K_{rk}(z, w),$$
$$|(Uw)_n| = \frac{|\langle z_k'', w_k''\rangle|}{|z_k''|} = |P_{z_k''} w|.$$

□

Consider the function $\Phi_n(\rho, u)$ defined by (2.38). It can be easily checked that

$$\Phi_n(\rho, u) = \frac{\Gamma(n + \rho/2)}{\Gamma(\rho/2)}(1-u)^{-n} Q_{n-1}(\rho, u), \tag{2.43}$$

where

$$Q_n(\rho, u) = \sum_{k=0}^{n} \frac{(-1)^k}{k+\rho/2}\binom{n}{k} u^k.$$

Set

$$\lambda_n(\rho) := \min\{\,|u| : Q_{n-1}(\rho, u) = 0\,\}.$$

Proposition 2.21.

(i) *If* $n \leq 4$, *then for every* $1 \leq \rho < \infty$, $\lambda_n(\rho) > 1$.

(ii) *For every* $n \geq 5$ *there exist* ρ, $1 \leq \rho < \infty$, *such that* $\lambda_n(\rho) < 1$.

(iii) *For all* $1 \leq \rho < \infty$

$$\lambda_n(\rho) \geq \frac{\rho + 2}{\rho + 2n}. \tag{2.44}$$

Proof. It is easy to check that $Q_n(\rho, u)$ has no zeros in the unit disk for $n \leq 2$. Consider the polynomial

$$Q_3(\rho, u) = \frac{1}{\rho/2} - \frac{3u}{1+\rho/2} + \frac{3u^2}{2+\rho/2} - \frac{u^3}{3+\rho/2}.$$

For $u \in R$

$$\frac{dQ_3(\rho, u)}{du} < 0.$$

Consequently $Q_3(\rho, u)$ has the only real zero. It can be easily proved that this zero belongs to the interval $(1, 1 + 6/\rho)$, and so the moduli of the other zeros are greater than 1.

An easy computation shows that

$$Q_n(\rho, u) = \int_0^1 t^{\rho/2-1}(1-ut)^n\, dt. \tag{2.45}$$

Set $s = 2 + \rho/2$. Then

$$Q_4(\rho, u) = \frac{1}{s-2} - \frac{4u}{s-1} + \frac{6u^2}{s} - \frac{4u^3}{s+1} + \frac{u^4}{s+2}.$$

It follows from (2.45) that $Q_4(\rho, u)$ has no real zeros. Let u_1, $\overline{u}_1$, u_2, and $\overline{u}_2$ be the zeros of $Q_4(\rho, u)$. In view of the Viete Theorem

$$\begin{cases} \operatorname{Re} u_1 + \operatorname{Re} u_2 = 2\dfrac{s+2}{s+1}, \\ |u_1|^2 + |u_2|^2 + 4 \operatorname{Re} u_1 \operatorname{Re} u_2 = 6\dfrac{s+2}{s}, \\ |u_1|^2 \operatorname{Re} u_2 + |u_2|^2 \operatorname{Re} u_1 = 2\dfrac{s+2}{s-1}, \\ |u_1|^2 |u_2|^2 = \dfrac{s+2}{s-2}. \end{cases}$$

Suppose $|u_1| = 1$. From the first, third, and fourth equations of this system we find

$$\operatorname{Re} u_1 = \frac{s^2-4}{s^2-1},$$
$$\operatorname{Re} u_2 = \frac{s^2+2s}{s^2-1}.$$

Using the second equation, we obtain $s^2 + 2 = 0$. This contradiction proves that $Q_4(\rho, u)$ has no zeros with modulus equals 1.

Let $\rho = 2$. Then from (2.45)

$$Q_4(2, u) = \frac{1-(1-u)^5}{5u}.$$

Hence $Q_4(2, u)$ has no zeros in the closed unit disk. Suppose that there exists a ρ_0 such that $Q_4(\rho_0, u)$ has zeros inside the unit disk. We can easily check that $Q_4(\rho, u)$ has no multiple zeros and hence the zeros of $Q_4(\rho, u)$ are differentiable functions of ρ. Therefore there exists a ρ_1 such that $Q_4(\rho_1, u)$ has a zero with modulus equal to 1 but we proved previously that this is impossible.

To prove (*ii*) note that from (2.45)

$$Q_n(2,u) = \frac{1-(1-u)^{n+1}}{(n+1)u}.$$

Thus $u_n = 1 - e^{2\pi i/(n+1)}$ is the zero of $Q_n(2,u)$. Since $|u_n| < 1$ for all $n \geq 6$, we need only to prove that $Q_5(\rho, u)$ has zeros inside the unit disk for some $\rho \geq 1$. Let $u(\rho)$ be the zero of $Q_5(\rho, u)$ such that $u(2) = u_5$. Then it can be checked that

$$\left.\frac{du(\rho)}{d\rho}\right|_{\rho=2} = \frac{119 + i56\sqrt{3}}{360}.$$

Thus for $\rho > 2$ and sufficiently close to 2, $|u(\rho)| < 1$.

To prove (*iii*) it suffices to show that the function $\Phi_n(\rho, u)$ does not vanish in the disk

$$|u| < \frac{\rho+2}{\rho+2n}.$$

Consider the function

$$\phi(v) := \Phi_n\left(\rho, \frac{\rho+2}{\rho+2n}v\right) = \sum_{k=0}^{\infty} d_k v^k.$$

It is easy to check that $d_0 = d_1 > d_2 > \ldots > d_k > \ldots > 0$. Thus

$$\begin{aligned}|(1-v)\phi(v)| &= \left|d_0 - \sum_{k=0}^{\infty}(d_k - d_{k+1})v^{k+1}\right| \\ &\geq d_0 - \left|\sum_{k=0}^{\infty}(d_k - d_{k+1})v^{k+1}\right| \geq d_0 - |v|^2 d_1 = (1-|v|^2)d_0.\end{aligned}$$

Hence $\phi(v) \neq 0$ if $|v| < 1$. □

Put

$$\Delta_{nk}(\rho) := \left\{ z \in B_n : \frac{|z_k''|^2}{\lambda_n^2(\rho)} + |z_k'|^2 < 1 \right\}, \quad \Delta_{nk}(\infty) := B_n.$$

It is clear that $\Delta_{nk}(\rho) = B_n$ if $\lambda_n(\rho) \geq 1$. In particular, by Proposition 2.21 it follows that for all $1 \leq p \leq \infty$ and $1 \leq n \leq 5$, $\Delta_{nk}(\rho) = B_n$.

Theorem 2.22. *Let $1 \le p \le \infty$ and $A = A_k := \{ z \in B_n : z_{n-k+1} = \ldots = z_n = 0 \}$, $1 \le k \le n$. For all $a \in \Delta_{nk}(rp)$ the method*

$$f(a) \approx \begin{cases} \chi^{(2-p)/p}(a) \sum_{j=0}^{r-1} \frac{1}{j!} d^j \left(\chi^{(p-2)/p}(z) f(z) \right)\Big|_{z=a'_k}, & 1 \le p < \infty, \\ (1 - |a|^2) \sum_{j=0}^{r-1} \frac{1}{j!} d^j \left(\frac{f(z)}{1 - \langle z, a \rangle} \right)\Big|_{z=a'_k}, & p = \infty, \end{cases}$$

where $dz = a''_k$ and

$$\chi(z) := \frac{\Phi_n(rp, s_k(z,a))}{(n-1)!(1 - \langle z, a'_k \rangle)^{n+rp/2}},$$

is optimal for the class $H_p(B_n)$. Moreover,

$$e(a, I^r_{A_k}, H_p(B_n)) = \begin{cases} \chi^{1/p}(a) |a''_k|^r, & 1 \le p < \infty, \\ \left(\frac{|a''_k|}{\sqrt{1 - |a'_k|^2}} \right)^r, & p = \infty. \end{cases}$$

Proof. For $a \in A_k$ the statement is evident. Assume that $|a''_k| \ne 0$. Let $1 \le p < \infty$. Since

$$\sup_{z \in B_n} |s_k(z,a)| = \frac{|a''_k|}{\sqrt{1 - |a'_k|^2}}, \tag{2.46}$$

for $a \in \Delta_{nk}(rp)$ the polynomial $Q_n(rp, s_k(z,a))$ does not vanish for all $z \in \overline{B}_n$. Thus it follows from (2.43) that $\chi^s(z) \in \mathcal{H}_\infty(B_n)$ for any $s \in R$. Set

$$g(z) := \chi^{2/p}(z) \langle z, a''_k \rangle^r, \quad \gamma := \chi^{(2-p)/p}(a) |a''_k|^{r(p-2)}.$$

We have

$$\begin{aligned} \overline{g(z)} |g(z)|^{p-2} &= \overline{\chi(z)} \chi^{(p-2)/p}(z) \langle a''_k, z \rangle^r |\langle a''_k, z \rangle|^{r(p-2)} \\ &= K_{rk}(a,z) \chi^{(p-2)/p}(z) |\langle a''_k, z \rangle|^{r(p-2)}. \end{aligned}$$

Taking into account Proposition 2.20 and the fact that $\chi^{(p-2)/p} f \in \mathcal{H}_p(B_n)$, for all $f \in \mathcal{H}_p(B_n)$ we have

$$\gamma \int_S \overline{g(z)} |g(z)|^{p-2} f(z)\, d\sigma(z)$$
$$= \chi^{(2-p)/p}(a) \int_S K_{rk}(a,z) \chi^{(p-2)/p}(z) f(z) |P_{a_k''} z|^{r(p-2)}\, d\sigma(z)$$
$$= f(a) - \chi^{(2-p)/p}(a) \sum_{j=0}^{r-1} \frac{1}{j!} d^j \left(\chi^{(p-2)/p}(z) f(z) \right)\Big|_{z=a_k'}, \qquad (2.47)$$

where $dz = a_k''$. It is easy to see that for all $z \in A_k$ $(D^\alpha g)(z) = 0$, $|\alpha| = 0, \ldots, r-1$. Thus the conditions of Theorem 1.10 are fulfilled. It remains to find $\|g\|_{\mathcal{H}_p(B_n)}$. Substituting $f = g$ into (2.47), we have

$$\gamma \|g\|^p_{\mathcal{H}_p(B_n)} = g(a).$$

Hence,

$$\|g\|_{\mathcal{H}_p(B_n)} = \chi^{1/p}(a) |a_k''|^r.$$

Consider now the case $p = \infty$. We reduce it to the one-dimensional problem of recovery. Let $f \in \mathcal{H}_\infty(B_n)$ and $b \in B_n$. Put $\varphi(u) := f(b_1, \ldots, b_{n-1}, \sqrt{1 - |b_1'|^2} u)$. Obviously, $\varphi \in H_\infty(B_1)$. From Theorem 2.1 for all $|\xi| < 1$ we obtain

$$|\varphi(\xi) - \sum_{j=0}^{r-1} c_j \varphi^{(j)}(0)| \le |\xi|^r,$$

where

$$c_j = \frac{\xi^r (1 - |\xi|^2)}{j!(r-j-1)!} \left[\frac{1}{(1 - \bar{\xi} u)(\xi - u)} \right]^{(r-j-1)}_{\Big|u=0}.$$

We have

$$\sum_{j-0}^{r-1} c_j \varphi^{(j)}(0) = \frac{\xi^r (1 - |\xi|^2)}{(r-1)!} \left[\frac{\varphi(u)}{(1 - \bar{\xi} u)(\xi - u)} \right]^{(r-1)}_{\Big|u=0}$$
$$= (1 - |\xi|^2) \sum_{j=0}^{r-1} \frac{\xi^j}{j!} \left(\frac{\varphi(u)}{1 - \bar{\xi} u} \right)^{(j)}_{\Big|u=0}.$$

Making the substitution $v=\sqrt{1-|b_1'|^2}u$ and putting $\xi=b_n(1-|b_1'|^2)^{-1/2}$, we get

$$\sum_{j=0}^{r-1} c_j\varphi^{(j)}(0)=(1-|b|^2)\sum_{j=0}^{r-1}\frac{b_n^j}{j!}D_n^j\left(\frac{f(z)}{1-\langle z,b\rangle}\right)\Big|_{z=b_1'}$$

$$=(1-|b|^2)\sum_{j=0}^{r-1}\frac{1}{j!}d^j\left(\frac{f(z)}{1-\langle z,b\rangle}\right)\Big|_{z=b_1'},$$

where $dz=b_n''$. Thus, for all $b\in B_n$

$$\left|f(b)-(1-|b|^2)\sum_{j=0}^{r-1}\frac{1}{j!}d^j\left(\frac{f(z)}{1-\langle z,b\rangle}\right)\Big|_{z=b_1'}\right|\le\left(\frac{|b_k''|}{\sqrt{1-|b_k'|^2}}\right)^r. \tag{2.48}$$

Let $a\in B\setminus A_k$. Consider the matrix U of the form (2.42) where C is the unitary $k\times k$ matrix which transforms $(a_{n-k+1},\ldots,a_n)$ into $(0,\ldots,0,|a_k''|)$. Set $b=Ua$. Then substituting $f(U^{-1}z)$ into (2.48), we have

$$\left|f(a)-(1-|Ua|^2)\sum_{j=0}^{r-1}\frac{1}{j!}d^j\left(\frac{f(U^{-1}z)}{1-\langle z,Ua\rangle}\right)\Big|_{z=(Ua)_1'}\right|$$

$$=\left|f(a)-(1-|a|^2)\sum_{j=0}^{r-1}\frac{1}{j!}d^j\left(\frac{(}{f(w)}1-\langle w,a\rangle\right)\Big|_{w=a_k'}\right|\le\left(\frac{|a_k''|}{\sqrt{1-|a_k'|^2}}\right)^r,$$

where $dz=a_k''$. Thus,

$$e(a,I_{A_k}^r,H_\infty(B_n))=\left(\frac{|a_k''|}{\sqrt{1-|a_k'|^2}}\right)^r.$$

On the other hand, in view of (2.46) the function

$$g_0(z)=\left(\frac{\sqrt{1-|a_k'|^2}}{|a_k''|}\frac{\langle z,a_k''\rangle}{1-\langle z,a_k'\rangle}\right)^r$$

lies in $H_\infty(B_n)$ and for all $z\in A_k$ $(D^\alpha g_0)(z)=0$, $|\alpha|=0,\ldots,r-1$. Consequently, from Theorem 1.6 it follows that

$$e(a,I_{A_k}^r,H_\infty(B_n))=\sup_{\substack{f\in H_\infty(B_n)\\ I_{A_k}^r f=0}}|f(a)|\ge|g_0(a)|=\left(\frac{|a_k''|}{\sqrt{1-|a_k'|^2}}\right)^r.$$

□

Since $e(a, I^r_{A_k}, H_p(B_n))$ coincides with the solution of the extremal problem

$$\sup_{\substack{f\in H_p(B_n)\\ I^r_{A_k} f=0}} |f(a)|$$

(see Theorem 1.6), for $k = n$ and $A_n = \{0\}$ we obtain the following generalization of the Schwartz Lemma.

Corollary 2.23. *For all $1 \le p < \infty$ and all $a \in B_n$ such that $|a| < \lambda_n(rp)$*

$$\sup_{\substack{f\in H_p(B_n)\\ (D^\alpha f)(0)=0,\ |\alpha|\le r-1}} |f(a)|$$

$$= \frac{|a|^r}{(1-|a|^2)^{n/p}} \left[\frac{\Gamma(n+rp/2)}{\Gamma(n)\Gamma(rp/2)} \sum_{k=0}^{n-1} \frac{(-1)^k}{k+rp/2} \binom{n-1}{k} |a|^{2k} \right]^{1/p}.$$

Note that if $n \ge 6$ we obtain the solution of recovery problem at a point $a \in B_n$ only for $a \in \Delta_{nk}(rp)$ (as it was mentioned above, for $1 \le n \le 5$, $\Delta_{nk}(rp) = B_n$ for all $1 \le p \le \infty$). Nevertheless, using (2.44), we can find a simple domain

$$\left(\frac{rp+2n}{rp+2}\right)^2 |a_k''|^2 + |a_k'|^2 < 1$$

contained in $\Delta_{nk}(rp)$. From the last inequality it is easy to see that our method can be applied for any $a \in B_n$ if r is sufficiently large.

Following Rudin [1980], we say that A is an affine subset of B_n, if it is the intersection of any affine subset of C^n with B_n. Let A be an arbitrary affine subset of B_n. In an appropriate coordinate system A has the following form

$$A = \{ z \in B_n \mid z_{n-k+1} = c_{n-k+1}, \dots, z_n = c_n \}, \tag{2.49}$$

where $c = (0, \dots, 0, c_{n-k+1}, \dots, c_n) \in B_n$, $1 \le k \le n$.

Put

$$\varphi_c(z) := \frac{c - P_c z - \sqrt{1-|c|^2}(z - P_c z)}{1 - \langle z, c\rangle}.$$

The mapping φ_c is an automorphism of B_n (see Rudin [1980]). Set

$$\Delta_{nk}(p, c) := \varphi_c(\Delta_{nk}(p)).$$

Theorem 2.24. *Let $1 \le p \le \infty$ and A is defined by (2.49). Then for all $a \in \Delta_{nk}(rp, c)$ the method*

$$f(a) \approx \begin{cases} \left(\dfrac{1-|c|^2}{1-\langle a, c\rangle}\right)^{2n/p} \chi_c^{(2-p)/p}(a_c) \\ \qquad \times \displaystyle\sum_{j=0}^{r-1} \frac{1}{j!} d^j \left[\frac{\chi_c^{(p-2)/p}(z) f(\varphi_c(z))}{(1-\langle z, c\rangle)^{2n/p}}\right]\Bigg|_{z=(a_c)'_k}, & 1 \le p < \infty, \\ (1-|a_c|^2) \displaystyle\sum_{j=0}^{r-1} \frac{1}{j!} d^j \left[\frac{f(\varphi_c(z))}{(1-\langle z, a_c\rangle)}\right]\Bigg|_{z=(a_c)'_k}, & p = \infty, \end{cases}$$

where $a_c = \varphi_c(a)$, $dz = (a_c)''_k$, and

$$\chi_c(z) := \frac{\Phi_n(rp, s_k(z, a_c))}{(n-1)!(1-\langle z, (a_c)'_k\rangle)^{n+rp/2}}$$

is optimal for the class $H_p(B_n)$. Moreover,

$$e(a, I_A^r, BH_p) = \begin{cases} \dfrac{(1-|c|^2)^{n/p}}{|1-\langle a, c\rangle|^{2n/p}} \chi^{1/p}(a_c)|(a_c)''_k|^r, & 1 \le p < \infty, \\ \left(\dfrac{|(a_c)''_k|}{\sqrt{1-|(a_c)'_k|^2}}\right)^r, & p = \infty. \end{cases}$$

Proof. First we recall some properties of φ_c (see Rudin [1980]):

$$\varphi_c(\varphi_c(z)) = z, \quad z \in B_n, \tag{2.50}$$

$$1 - \langle \varphi_c(z), \varphi_c(w)\rangle = \frac{(1-|c|^2)(1-\langle z, w\rangle)}{(1-\langle z, c\rangle)(1-\langle c, w\rangle)}, \quad z, w \in B_n. \tag{2.51}$$

Moreover, the operator

$$(Tf)(z) := \frac{(1-|c|^2)^{n/p}}{(1-\langle z, c\rangle)^{2n/p}} f(\varphi_c(z))$$

is an isometry of $\mathcal{H}_p(B_n)$, that is, for all $f \in \mathcal{H}_p(B_n)$

$$\|Tf\|_{\mathcal{H}_p(B_n)} = \|f\|_{\mathcal{H}_p(B_n)}.$$

Let $1 \le p < \infty$. In view of (2.50) we have

$$a_c = \varphi_c(a) \in \varphi_c(\Delta_{nk}(rp, c)) = \Delta_{nk}(rp).$$

Consider an arbitrary function $f \in H_p(B_n)$. Then $g := Tf \in \mathcal{H}_p(B_n)$. By Theorem 2.22 we obtain

$$\left|g(a_c) - \chi_c^{(2-p)/p}(a_c)\sum_{j=0}^{r-1}\frac{1}{j!}d^j\left(\chi_c^{(p-2)/p}(z)g(z)\right)\Big|_{z=(a_c)'_k}\right| \le e(a_c, I^r_{A_k}, H_p(B_n)) \quad (2.52)$$

(here and further $dz = (a_c)''_k$). From (2.50) and (2.51)

$$g(a_c) = \frac{(1-|c|^2)^{n/p}}{(1-\langle a_c, c\rangle)^{2n/p}}f(a) = \frac{(1-|c|^2)^{n/p}}{(1-\langle \varphi_c(a), \varphi_c(0)\rangle)^{2n/p}}f(a) = \frac{(1-\langle a, c\rangle)^{2n/p}}{(1-|c|^2)^{n/p}}f(a).$$

Multiplying both sides of (2.52) by

$$\frac{(1-|c|^2)^{n/p}}{|1-\langle a, c\rangle|^{2n/p}},$$

we have

$$\left|f(a) - \left(\frac{1-|c|^2}{1-\langle a, c\rangle}\right)^{2n/p}\chi_c^{(2-p)/p}(a_c) \times\sum_{j=0}^{r-1}\frac{1}{j!}d^j\left[\frac{\chi_c^{(p-2)/p}(z)f(\varphi_c(z))}{(1-\langle z, c\rangle)^{2n/p}}\right]\Big|_{z=(a_c)'_k}\right| \le \frac{(1-|c|^2)^{n/p}}{|1-\langle a, c\rangle|^{2n/p}}e(a_c, I^r_{A_k}, H_p(B_n)).$$

It is easy to check that $\varphi_c(A) = A_k$ and consequently $\varphi_c(A_k) = A$. In view of arbitrariness of f we have

$$e(a, I^r_A, H_p(B_n)) \le \frac{(1-|c|^2)^{n/p}}{|1-\langle a, c\rangle|^{2n/p}}e(a_c, I^r_{A_k}, H_p(B_n)).$$

Consider the function

$$f_0(z) - \chi_c^{-1/p}(a_c)|(a_c)''_k|^{-r}\chi_c^{2/p}(z)\langle z, (a_c)''_k\rangle^r$$

which is an extremal function in the problem of optimal recovery at the point a_c using the information $I^r_{A_k}$, i.e., $f_0 \in H_p(B_n)$ and $f_0(a_c) =$

$e(a_c, I^r_{A_k}, H_p(B_n))$. Set $g_0 := Tf_0$. Then $g_0 \in H_p(B_n)$, $I^r_A g_0 = 0$, and cosequently

$$e(a, I^r_A, H_p(B_n)) \geq |g_0(a)| = \frac{(1-|c|^2)^{n/p}}{|1-\langle a,c\rangle|^{2n/p}} e(a_c, I^r_{A_k}, H_p(B_n)).$$

The case $p = \infty$ is considered by the same scheme. □

Now we consider the analogous problem of optimal recovery for the Bergman space $\mathcal{A}_p(B_n)$ which is defined as the set of all holomorphic functions in B_n satisfying the condition

$$\|f\|_{\mathcal{A}_p(B_n)} := \left(\int_{B_n} |f(z)|^p \, d\nu(z)\right)^{1/p} < \infty, \quad 1 \leq p < \infty,$$

where ν is the normalized Lebesgue measure in B_n. Set

$$A_p(B_n) := \{ f \in \mathcal{A}_p(B_n) : \|f\|_{\mathcal{A}_p(B_n)} \leq 1 \}.$$

Since $A_\infty = H_\infty$ we study only the case $1 \leq p < \infty$.

We construct an optimal method of recovery of $f \in A_p(B_n)$ at a point $a \in B_n$ from the information I^r_A where A is defined by (2.49). We also find the intrinsic error of this problem which we denote by $e(a, I^r_A, A_p(B_n))$.

Set

$$\widetilde{\Delta}_{nk}(p) := \left\{ z \in B_n : \frac{|z''_k|^2}{\lambda^2_{n+1}(p)} + |z'_k| < 1 \right\},$$

$$\widetilde{\Delta}_{nk}(p, c) := \varphi_c(\widetilde{\Delta}_{nk}(p)).$$

Theorem 2.25. *Let $1 \leq p < \infty$. Then for all $a \in \widetilde{\Delta}_{nk}(rp, c)$ the method*

$$f(a) \approx \left(\frac{1-|c|^2}{1-\langle a,c\rangle}\right)^{2(n+1)/p} \widetilde{\chi}_c^{(2-p)/p}(a_c) \times \sum_{j=0}^{r-1} \frac{1}{j!} d^j \left[\frac{\widetilde{\chi}_c^{(p-2)/p}(z) f(\varphi_c(z))}{(1-\langle z,c\rangle)^{2(n+1)/p}} \right]_{z=(a_c)'_k},$$

where $a_c = \varphi_c(a)$, $dz = (a_c)''_k$, and

$$\widetilde{\chi}_c(z) := \frac{\Phi_{n+1}(rp, s_k(z, a_c))}{n!(1-\langle z, (a_c)'_k\rangle)^{n+1+rp/2}},$$

is optimal for the class $A_p(B_n)$. Moreover,

$$e(a, I^r_A, A_p(B_n)) = \frac{(1-|c|^2)^{(n+1)/p}}{|1-\langle a,c\rangle|^{2(n+1)/p}} \widetilde{\chi}^{1/p}(a_c) |(a_c)''_k|^r.$$

Proof. Define the operator of embedding E by

$$(Eg)(z_0, z) := g(z),$$

where $z \in B_n$ and $(z_0, z) \in B_{n+1}$. It is known (see Rudin [1980]) that E is a linear isometric embedding of $\mathcal{A}_p(B_n)$ into $\mathcal{H}_p(B_{n+1})$, that is, for all $g \in \mathcal{A}_p(B_n)$, $Eg \in \mathcal{H}_p(B_{n+1})$ and $\|g\|_{\mathcal{A}_p(B_n)} = \|Eg\|_{\mathcal{H}_p(B_{n+1})}$. Let $e\colon C^n \to C^{n+1}$ be the embedding $ez := (0, z)$. Consider the problem of optimal recovery of functions $f \in H_p(B_{n+1})$ at a point ea from the information I^r_{eA}. It is easy to check that

$$e(\varphi_c(z)) = \varphi_{ec}(ez), \quad e(\widetilde{\Delta}_{nk}(p)) \subset \widetilde{\Delta}_{n+1,k}(p).$$

Since $a \in \widetilde{\Delta}_{nk}(rp, c)$ we have

$$\begin{aligned} ea \in e(\widetilde{\Delta}_{nk}(rp, c)) &= e\big(\varphi_c(\widetilde{\Delta}_{nk}(rp))\big) \\ &= \varphi_{ec}\big(e(\widetilde{\Delta}_{nk}(rp))\big) \subset \varphi_{ec}(\Delta_{n+1,k}(rp)) = \Delta_{n+1,k}(rp, ec). \end{aligned}$$

Let g be an arbitrary function from $A_p(B_n)$. Then $Eg \in H_p(B_{n+1})$. Using Theorem 2.24, we obtain

$$|(Eg)(ea) - T_0 I^r_{eA} Eg| \le e(ea, I^r_{eA}, H_p(B_{n+1})), \tag{2.53}$$

where T_0 is the appropriate optimal method of recovery. For any multiindex $\alpha = (\alpha_0, \dots, \alpha_n)$, $f \in \mathcal{A}_p(B_n)$, and $z \in B_n$

$$D^\alpha(Ef)\big|_{ez} = D^\alpha f\big|_z.$$

Thus (2.53) can be rewritten in the form

$$\begin{aligned} \Bigg| f(a) - \left(\frac{1-|c|^2}{1-\langle a, c\rangle}\right)^{2(n+1)/p} & \widetilde{\chi}_c^{(2-p)/p}(a_c) \\ \times \sum_{j=0}^{r-1} \frac{1}{j!} d^j & \left[\frac{\widetilde{\chi}_c^{(p-2)/p}(z) f(\varphi_c(z))}{(1-\langle z, c\rangle)^{2(n+1)/p}}\right]_{\Big| z=(a_c)'_k} \Bigg| \\ & \le \frac{(1-|c|^2)^{(n+1)/p}}{|1-\langle a, c\rangle|^{2(n+1)/p}} \widetilde{\chi}_c^{1/p}(a_c) |(a_c)''_k|^r, \end{aligned}$$

where $dz = (a_c)''_k$. Consequently,

$$e(a, I^r_A, A_p(B_n)) \le \frac{(1-|c|^2)^{(n+1)/p}}{|1-\langle a, c\rangle|^{2(n+1)/p}} \widetilde{\chi}^{1/p}(a_c) |(a_c)''_k|^r.$$

On the other hand, the function Eg_0, where

$$g_0(z) = \frac{(1-|c|^2)^{(n+1)/p}}{(1-\langle z,c\rangle)^{2(n+1)/p}} \frac{\widetilde{\chi}_c^{2/p}(\varphi_c(z))\langle \varphi_c(z),(a_c)_k''\rangle^r}{\widetilde{\chi}_c^{1/p}(a_c)|(a_c)_k''|^r},$$

coincides with an extremal function for the problem of optimal recovery at the point ea on the class $H_p(B_{n+1})$ using the information I_{eA}^r. Since $\|g_0\|_{A_p(B_n)} = \|Eg_0\|_{H_p(B_{n+1})} = 1$ and $I_A^r g_0 = 0$ we have

$$e(a, I_A^r, A_p(B_n)) \geq |g_0(a)| = \frac{(1-|c|^2)^{(n+1)/p}}{|1-\langle a,c\rangle|^{2(n+1)/p}} \widetilde{\chi}^{1/p}(a_c)|(a_c)_k''|^r.$$

□

Analogously to Corollary 2.23 we obtain the generalization of the Schwartz Lemma for the class $A_p(B_n)$.

Corollary 2.26. *For all $1 \leq p < \infty$ and $a \in B_n$ such that $|a| < \lambda_{n+1}(rp)$*

$$\sup_{\substack{f\in A_p(B_n)\\ (D^\alpha f)(0)=0,\, |\alpha|\leq r-1}} |f(a)|$$

$$= \frac{|a|^r}{(1-|a|^2)^{(n+1)/p}} \left[\frac{\Gamma(n+1+rp/2)}{\Gamma(n+1)\Gamma(rp/2)} \sum_{k=0}^{n} \frac{(-1)^k}{k+rp/2} \binom{n}{k} |a|^{2k}\right]^{1/p}.$$

Let us now consider the problem of optimal recovery for the Bergman class in the case $n = 1$. Using the easily checking equalities

$$\sum_{j=0}^{r-1} \frac{1}{j!} g^{(j)}(0) u^j = \frac{u^r}{(r-1)!} \left(\frac{g(z)}{u-z}\right)^{(r-1)}\Bigg|_{z=0},$$

$$F^{(r-1)}(0) = (-1)^{r-1}(1-|c|^2) \left[(1-\overline{c}t)^{r-2} F\left(\frac{c-t}{1-\overline{c}t}\right)\right]^{(r-1)}_{t=c},$$

we have from Theorem 2.25

Corollary 2.27. *Let $1 \leq p < \infty$, $I_c^r f := \{f(c), \ldots, f^{(r-1)}(c)\}$, $c \in B_1$. For all $a \in B_1$ the method*

$$f(a) \approx \frac{(a-c)^r (1-|a|^2)^{2-4/p}}{(1-\overline{c}a)^{r+1}\omega^{(p-2)/p}(a)} \frac{1}{(r-1)!} \left[\frac{(1-\overline{c}t)^{r+1}\omega^{(p-2)/p}(t) f(t)}{(t-a)(1-\overline{a}t)^{2-4/p}}\right]^{(r-1)}_{t=c},$$

where

$$\omega(t) := 1 + \frac{rp}{2}\left(1 - \frac{\bar{c}-\bar{a}}{1-c\bar{a}}\frac{c-t}{1-\bar{c}t}\right),$$

is optimal for the class $A_p(B_1)$. Moreover,

$$e(a, I_c^r, A_p(B_1)) = \left|\frac{c-a}{1-\bar{c}a}\right|^r \frac{\omega^{1/p}(a)}{(1-|a|^2)^{2/p}}.$$

2.6 Notes and References

2.1. Optimal method of recovery for H_∞ was obtained by Osipenko [1972], [1976]. The general case $1 \le p \le \infty$ was considered by Fisher, Micchelli [1980]. Various results connected with the problem of optimal recovery of functions from H_∞ by inaccurate data may be found in Osipenko [1982], [1985].

2.2. Optimal methods of recovery for functions which are analytic and bounded in an annulus were found by Ovchintsev [1989a] and Wilderotter [1992b]. Using the map $w = i^{-1}\log z$ this problem may be reduced to the equivalent one of optimal recovery on the class $H_\infty^\beta(T)$. Theorem 2.3 was proved by Osipenko, Wilderotter [1997]. Similar results for the Smirnov classes E_p of functions defined in an annulus were obtained by Ovchintsev [1989b].

2.3. The material of this section is taken from Osipenko [1994c]. Lemma 2.5 in a less general form was proved in Osipenko, Stessin [1993].

2.4. For $q = \infty$ and $E = [a,b] \subset (-1,1)$ the extremal problem (2.23) was solved by Osipenko [1972]. This problem is very close to the third problem of Zolotarev. Using the results of Gonchar [1969] on the generalized Zolotarev problem, Osipenko [1976] proved that

$$\lim_{n\to\infty} \sqrt[n]{s_n(H_\infty, C(E))} = e^{-\frac{1}{c(E,\mathbb{C}\setminus D)}},$$

where $E \subset D$ is a compact with a connected complement and $c(E,F)$ is the capacity of the condenser (E,F). For the Kolmogorov n-widths analogous results were obtained by Erokhin [1968] and Widom [1972] (see also Fisher, Micchelli [1980]). Some results connected with the minimization of Blaschke products in integral metrics were studied in Osipenko [1990].

Proposition 2.11 is due to Parfenov [1985]. The proof of Theorem 2.12 is based on ideas which were proposed by Babenko [1958] and Taikov [1967] (see comments to Corollary 4.7 in Section 4.7). Theorem 2.14 was proved in Osipenko [1994c]. In the same paper the problem (2.24) was solved for

$q = \infty$. Theorem 2.15 was obtained by Osipenko [1995a]. The equality (2.36) was proved in Osipenko, Wilderotter [1997].

2.5. For $r = 1$ (that is, the information operator has the form $If = f_{|A}$) the results of this section were obtained by Osipenko, Stessin [1992a]. The problem of optimal recovery for the Bergman classes in the one-dimensional case (Corollary 2.27) was considered in Osipenko, Stessin [1991]. Note that the appropriate extremal problem plays an essential role in finding canonical divisors in Bergman spaces (see Duren, Khavinson, Shapiro, Sundberg [1993]).

Chapter 3

Optimal Recovery of Derivatives

We consider here almost the same problems as in Chapter 2 for derivatives. In Section 3.1 we construct an optimal method of recovery of the first derivative while the general case is studying in Section 3.2. Section 3.3 is devoted to the periodic case. In Section 3.4 we find optimal methods of recovery of the first and second derivatives when their values are known with an error in the uniform norm. The problem of optimal recovery of derivatives from incorrect information are closely connected with the extremal problems which are known for the smooth functions as the Kolmogorov inequalities. In Section 3.5 we obtain the solution of these problems for analytic functions.

3.1 Optimal Recovery of the First Derivative

Consider now the problem of optimal recovery of the derivative of $f \in H_p$ at some point ξ from the unit disk D using the information operator

$$If := \{ f(z_1), \ldots , f^{(\nu_1-1)}(z_1), \ldots , f(z_n), \ldots , f^{(\nu_n-1)}(z_n) \},$$

where $z_1, \ldots , z_n$ are distinct points from D. Put

$$e_1(\xi, I, H_p) := \inf_{\varphi\colon C^N \to C} \sup_{f \in H_p} |f'(\xi) - \varphi(If)|.$$

Set

$$W(z) := \prod_{j=1}^{n} \left(\frac{z - z_j}{1 - \overline{z}_j z} \right)^{\nu_j}, \qquad \omega_j(z) := \prod_{\substack{s=1 \\ s \neq j}}^{n} \left(\frac{z - z_s}{1 - \overline{z}_s z} \right)^{\nu_s}.$$

We introduce the notation

$$\beta(\xi) := \frac{1-|\xi|^2}{2}W'(\xi) + \frac{\overline{\xi}}{p}W(\xi),$$

$$D_1 := \left\{ \xi \in D : |\beta(\xi)| < \frac{p-1}{p}|W(\xi)| \right\}, \qquad D_0 := D \setminus D_1,$$

$$b := \begin{cases} \dfrac{p}{p-1}\dfrac{\beta(\xi)}{W(\xi)}, & \xi \in D_1, \\[2ex] \dfrac{\overline{W(\xi)}e^{i\arg\beta(\xi)}}{|\beta(\xi)| + \sqrt{|\beta(\xi)|^2 - \dfrac{p-2}{p}|W(\xi)|^2}}, & \xi \in D_0, \end{cases}$$

$$a := \frac{\xi - \overline{b}}{1 - \overline{\xi b}}, \qquad u_\xi(z) := \begin{cases} 1, & \xi \in D_1, \\ \dfrac{z-a}{1-\overline{a}z}, & \xi \in D_0. \end{cases}$$

Theorem 3.1. *For all $1 \le p \le \infty$ the method*

$$f'(\xi) \approx \sum_{j=1}^{n} \sum_{\nu=0}^{\nu_j-1} c_{j\nu}(\xi,p) f^{(\nu)}(z_j),$$

where for $\xi \neq z_j$

$$c_{j\nu}(\xi,p) = -\frac{\alpha(\xi)}{\nu!(\nu_j-\nu-1)!}\left(\frac{u_\xi(z)(1-\overline{z}_j z)^{\nu_j}(1-\overline{a}z)^{2(p-1)/p}}{\omega_j(z)(z-\xi)^2(1-\overline{\xi}z)^{2(p-2)/p}} \right)^{(\nu_j-\nu-1)}\Bigg|_{z=z_j},$$

$$c_{j0}(z_j,p) = \frac{\omega_j'(z_j)}{\omega_j(z_j)} + \frac{2}{p}\frac{\overline{z}_j}{1-|z_j|^2},$$

and

$$\alpha(\xi) := \begin{cases} \dfrac{W(\xi)(1-|\xi|^2)^{2(p-2)/p}}{u_\xi(\xi)(1-\overline{a}\xi)^{2(p-1)/p}}, & \xi \notin \{z_1,\dots,z_n\}, \\[2ex] \dfrac{\omega_j(\xi)}{(1-|\xi|^2)^{2/p}}, & \xi = z_j,\ j=1,\dots,n, \end{cases}$$

is optimal for the class H_p. Moreover,

$$e_1(\xi, I, H_p) = \begin{cases} \dfrac{|W(\xi)|(1+|b|^2)^{\frac{p-1}{p}}}{|u_\xi(\xi)|(1-|\xi|^2)^{\frac{p+1}{p}}}, & \xi \notin \{z_1,\dots,z_n\}, \\[2ex] \dfrac{|\omega_j(\xi)|}{(1-|\xi|^2)^{\frac{p+1}{p}}}, & \xi = z_j,\ j=1,\dots,n. \end{cases}$$

Proof. Put

$$g(z) := \frac{z-a}{1-\overline{a}z}\,\frac{W(z)(1-\overline{a}z)^{2/p}}{u_\xi(\xi)(1-\overline{\xi}z)^{4/p}}.$$

Let $\xi \notin \{z_1, \dots, z_n\}$ and $1 \le p < \infty$. Then by the residue theorem for every $f \in \mathcal{H}_p$ we have

$$\begin{aligned}\alpha(\xi)\int_0^{2\pi} &\overline{g(z)}|g(z)|^{p-2} f(z)\,d\mu(z)\\ &= \frac{\alpha(\xi)}{2\pi i}\int_{|z|=1} \frac{u_\xi(z)(1-\overline{a}z)^{2(p-1)/p} f(z)\,dz}{W(z)(z-\xi)^2(1-\overline{\xi}z)^{2(p-2)/p}}\\ &= f'(\xi) + \lambda(a) f(\xi) - \sum_{j=1}^{n}\sum_{\nu=0}^{\nu_j-1} c_{j\nu}(\xi,p) f^{(\nu)}(z_j), \qquad (3.1)\end{aligned}$$

where

$$\lambda(a) = \alpha(\xi)\left(\frac{u_\xi(z)(1-\overline{a}z)^{2(p-1)/p}}{W(z)(1-\overline{\xi}z)^{2(p-2)/p}}\right)'\Bigg|_{z=\xi}.$$

It is easy to check that a is chosen so that $\lambda(a) = 0$. Substituting $f(z) = g(z)$ in (3.1), we get

$$\|g\|^p_{\mathcal{H}_p} = \frac{1}{2\pi i}\int_{|z|=1} \frac{(1-\overline{a}z)(z-a)\,dz}{(z-\xi)^2(1-\overline{\xi}z)^2} = \frac{1+|b|^2}{(1-|\xi|^2)|1-\xi b|^2}.$$

Now the required statement for the considered case follows from Theorem 1.10. For $\xi = z_k$, $a = \xi$, and analogously to the previous case we have

$$\begin{aligned}\alpha(\xi)\int_0^{2\pi} \overline{g(z)}|g(z)|^{p-2} f(z)\,d\mu(z) &= \frac{\alpha(\xi)}{2\pi i}\int_{|z|=1} \frac{(1-\overline{\xi}z)^{2/p} f(z)\,dz}{\omega_k(z)(z-\xi)^2}\\ &= f'(\xi) - \sum_{j=1}^{n}\sum_{\nu=0}^{\nu_j-1} c_{j\nu}(\xi,p) f^{(\nu)}(z_j)\end{aligned}$$

(it is necessary to take into account that $\nu_k = 1$ in this case). Then we use Theorem 1.10.

For $p = \infty$ we use the same scheme and the equality

$$\alpha(\xi)\int_0^{2\pi} \overline{g(z)}|\varphi(z)| f(z)\,d\mu(z) = \frac{\alpha(\xi)}{2\pi i}\int_{|z|=1} \frac{u_\xi(z)(1-\overline{a}z)^2 f(z)\,dz}{W(z)(z-\xi)^2(1-\overline{\xi}z)^2},$$

where

$$\varphi(z) = \frac{(1-\overline{a}z)^2}{(1-\overline{\xi}z)^4}.$$

□

From Theorem 3.1 it follows that in a general situation the unit disk D is divided into two parts D_0 and D_1 such that for $\xi \in D_0$ the extremal function $g/\|g\|_{\mathcal{H}_p}$ has zeros with corresponding multiplicities only at the points $z_1, \dots, z_n$ and for $\xi \in D_1$ there is the additional zero at the point a. The appearance of such two sets was apparently first discovered by Dieudonné [1931]. He solved the extremal problem

$$\sup_{\substack{f\in H_\infty \\ f(0)=0}} |f'(\xi)|$$

and found that for $|\xi| \le \sqrt{2}-1$, $f(z) = z$ is an extremal function and for $\sqrt{2}-1 < |\xi| < 1$ an extremal function has the additional zero at some point $a(\xi)$.

Consider the generalized Dieudonné problem

$$\sup_{\substack{f\in H_p \\ f(0)=\dots=f^{(n-1)}(0)=0}} |f'(\xi)|.$$

In this case the set D_0 has the form

$$D_0 = \left\{\xi \in D : \left(n - \frac{2}{p}\right)|\xi|^2 + 2\frac{p-1}{p}|\xi| - n \le 0\right\}.$$

It is easy to check that for $1 \le p \le 2$, $D_0 = D$ and for $2 < p \le \infty$, $D_0 = \{\xi \in C : |\xi| \le r_{np}\}$, where

$$r_{np} = \frac{\sqrt{\left(\frac{p-1}{p}\right)^2 + n\left(n - \frac{2}{p}\right)} - \frac{p-1}{p}}{n - \frac{2}{p}}.$$

From Theorem 3.1 it also follows the solution of the problem of optimal recovery of $f'(0)$, $f \in H_p$, from the information $If = \{f(-r), f(-r)\}$, $r \in (0,1)$. For all $1 \le p \le \infty$ an optimal method of recovery has the form

$$f'(0) \approx (1-r^2)\frac{f(r)-f(-r)}{2r}.$$

3.2 A General Case

Let us consider now a rather general case. Namely we shall study the problem of optimal recovery of the linear functional

$$L_\xi^\lambda f := \sum_{j=\nu}^{\nu+m} \frac{\lambda_j}{j!} f^{(j)}(\xi), \tag{3.2}$$

where $\lambda_j \in C$, $\lambda_{\nu+m} \neq 0$, $\nu \geq 0$, on the class H_p on the basis of the information operator

$$If := \{ f(\xi), \dots, f^{(\nu-1)}(\xi), f(z_1), \dots, f^{(\nu_1-1)}(z_1), \dots, \\ f(z_n), \dots, f^{(\nu_n-1)}(z_n) \}$$

($\xi, z_1, \dots, z_n$ are distinct points from the unit disk D). The appropriate intrinsic error we denote by

$$e(\xi, \lambda, I, H_p) := \inf_{\varphi :\, C^N \to C} \sup_{f \in H_p} |L_\xi^\lambda f - \varphi(If)|.$$

In view of duality (see Theorem 1.6) we have

$$e(\xi, \lambda, I, H_p) = \sup_{\substack{f \in H_p \\ If = 0}} |L_\xi^\lambda f|. \tag{3.3}$$

In the previous section we deal with the case $m = 1$, $\nu = 0$, and $\lambda_0 = 1$.

To solve this general recovery problem, we need first to consider the Carathéodory–Fejér problem about finding the value

$$\inf \|f\|_{\mathcal{H}_p},$$

where the infimum is taking over all functions from $\mathcal{H}_p$ satisfying the condition

$$f^{(j)}(\xi) = j! c_j, \quad j = 0, \dots, m \tag{3.4}$$

($\xi \in D$ and $c_0, \dots, c_m$ are fixed complex numbers).

It is known (Khavinson [1981], see also Macintyer, Rogosinski [1950] and Rogosinski, Shapiro [1953]) that for all $1 \leq p \leq \infty$ among all functions satisfying (3.4) there exists a unique function of the kind

$$f_0(z) = C(1 - \bar{\xi} z)^{-2(m+1)/p} \prod_{j=1}^{k} \frac{z - \alpha_j}{1 - \bar{\alpha}_j z} \prod_{j=1}^{m} (1 - \bar{\alpha}_j z)^{2/p}, \tag{3.5}$$

where C is some constant, $0 \le k \le m$, $|\alpha_j| < 1$, $j = 1, \dots, k$, and $|\alpha_j| \le 1$, $j = k+1, \dots, m$. Moreover, this function is the unique solution of the Carathéodory–Fejér problem with the condition (3.4). (As usually all expressions involving p with $p = \infty$ are understood as their limit values as $p \to \infty$).

Without loss of generality assume that $c_0 \neq 0$ otherwise the problem reduces to the one with the number of parameters less than $m+1$. Set

$$d_j := \frac{c_j}{c_0}, \quad j = 0, \dots, m,$$

$$A := \begin{pmatrix} 1 & 0 & \dots & 0 \\ \frac{d_1}{2} & \frac{1}{2} & \dots & 0 \\ \dots & \dots & \dots & \dots \\ \frac{d_{m-1}}{m} & \frac{d_{m-2}}{m} & \dots & \frac{1}{m} \end{pmatrix}, \quad \begin{pmatrix} \rho_1 \\ \rho_2 \\ \vdots \\ \rho_m \end{pmatrix} := A^{-1} \begin{pmatrix} d_1 \\ d_2 \\ \vdots \\ d_m \end{pmatrix}, \tag{3.6}$$

$$\gamma_s := (-1)^{s+1} \left[\frac{2}{p} \bar{\xi}^s + \sum_{j=1}^{s} (-1)^j \binom{s}{j} \bar{\xi}^{s-j} (1 - |\xi|^2)^j \rho_j \right], \quad s = 1, \dots, m.$$

Theorem 3.2. *Let $1 \le p \le \infty$. The function (3.5) is the solution of the Carathéodory–Fejér problem with the condition (3.4) and $c_0 \neq 0$ if and only if the numbers*

$$b_j = \frac{\bar{\alpha}_j - \bar{\xi}}{1 - \xi \bar{\alpha}_j} \tag{3.7}$$

are the solution of the system

$$\sum_{j=1}^{k} (\bar{b}_j^{-s} - b_j^s) + \frac{2}{p} \sum_{j=1}^{m} b_j^s = \gamma_s, \quad s = 1, \dots, m, \tag{3.8}$$

such that

$$\begin{array}{ll} |b_j| < 1, & j = 1, \dots, k, \\ |b_j| \le 1, & j = k+1, \dots, m, \end{array} \quad 0 \le k \le m, \tag{3.9}$$

and

$$C = c_0 (1 - |\xi|^2)^{2(m+1)/p} \prod_{j=1}^{k} \frac{1 - \bar{\alpha}_j \xi}{\xi - \alpha_j} \prod_{j=1}^{m} (1 - \bar{\alpha}_j \xi)^{-2/p}. \tag{3.10}$$

Proof. Let the function (3.5) be the solution of the Carathéodory–Fejér problem. The equality (3.10) immediately follows from the fact that $f_0(\xi) = c_0$. Let us prove (3.8) for b_j defined by (3.7). Set

$$\begin{aligned} x(z) &:= \frac{f_0'(z)}{f_0(z)} \\ &= \sum_{j=1}^{k}\left(\frac{1}{z-\alpha_j} + \frac{\overline{\alpha}_j}{1-\overline{\alpha}_j z}\right) - \frac{2}{p}\sum_{j=1}^{m}\frac{\overline{\alpha}_j}{1-\overline{\alpha}_j z} + (m+1)\frac{2}{p}\frac{\overline{\xi}}{1-\overline{\xi}z}. \end{aligned} \tag{3.11}$$

From the definition of $x(z)$ it follows that

$$f_0^{(r+1)}(\xi) = \sum_{j=0}^{r}\binom{r}{j} f_0^{(r-j)}(\xi)x^{(j)}(\xi). \tag{3.12}$$

Taking into account (3.4), we obtain

$$\frac{1}{r+1}\sum_{j=0}^{r} d_{r-j}\frac{x^{(j)}(\xi)}{j!} = d_{r+1}, \quad r = 0, \dots, m-1.$$

Thus

$$\frac{x^{(j)}(\xi)}{j!} = \rho_{j+1}, \quad j = 0, \dots, m-1.$$

On the other hand, from (3.11) we have

$$\begin{aligned} \frac{x^{(r-1)}(\xi)}{(r-1)!} &= \sum_{j=1}^{k}\left(\frac{(-1)^{(r-1)}}{(\xi-\alpha_j)^r} + \frac{\overline{\alpha}_j^r}{(1-\overline{\alpha}_j\xi)^r}\right) \\ &\quad - \frac{2}{p}\sum_{j=1}^{m}\frac{\overline{\alpha}_j^r}{(1-\overline{\alpha}_j\xi)^r} + (m+1)\frac{2}{p}\frac{\overline{\xi}^r}{(1-|\xi|^2)^r}. \end{aligned}$$

In view of (3.7)

$$\frac{1}{\xi-\alpha_j} = -\frac{\overline{b_j}^{-1}+\overline{\xi}}{1-|\xi|^2}, \quad \frac{\overline{\alpha}_j}{1-\overline{\alpha}_j\xi} = \frac{b_j+\overline{\xi}}{1-|\xi|^2}.$$

Consequently,

$$\begin{aligned} \omega_r &:= \sum_{j=1}^{k}\left[(\overline{b_j}^{-1}+\overline{\xi})^r - (b_j+\overline{\xi})^r\right] + \frac{2}{p}\sum_{j=1}^{m}(b_j+\overline{\xi})^r \\ &= (m+1)\frac{2}{p}\overline{\xi}^r - (1-|\xi|^2)^r\rho_r, \quad r = 1, \dots, m. \end{aligned}$$

Hence

$$\sum_{j=1}^{k}(\overline{b}_j^{-s} - b_j^s) + \frac{2}{p}\sum_{j=1}^{m} b_j^s$$

$$= \sum_{j=1}^{k}\left[(\overline{b}_j^{-1} + \overline{\xi} - \overline{\xi})^s - (b_j + \overline{\xi} - \overline{\xi})^s\right] + \frac{2}{p}\sum_{j=1}^{m}(b_j + \overline{\xi} - \overline{\xi})^s$$

$$= \sum_{r=1}^{s}\binom{s}{r}(-\overline{\xi})^{s-r}\omega_r + m\frac{2}{p}(-\overline{\xi})^s = \gamma_s. \quad (3.13)$$

Since $|\alpha_j| < 1$, $j = 1, \dots, k$, and $|\alpha_j| \le 1$, $j = k+1, \dots, m$, the same inequalities hold for $b_1, \dots, b_m$.

Assume now that $b_1, \dots, b_m$ are solutions of (3.8) satisfying (3.9). Define α_j from (3.7) and consider the function (3.5). Using the same arguments in the reverse order, we get from (3.12)

$$\frac{f_0^{(j)}(\xi)}{f_0(\xi)} = d_j = \frac{c_j}{c_0}, \quad j = 0, \dots, m.$$

Choosing C from the condition $f_0(\xi) = c_0$ (it means that C is defined by (3.10)), we obtain that the equalities (3.4) are fulfilled. □

Corollary 3.3. *For all* $1 \le p \le \infty$ *and any* $\gamma_1, \dots, \gamma_m \in C$ *there exists a* k, $0 \le k \le m$, *for which the system* (3.8) *has a solution satisfying* (3.9). *Moreover, the function*

$$\prod_{j=1}^{k}\frac{z - \overline{b}_j}{1 - b_j z}\prod_{j=1}^{m}(1 - b_j z)^{2/p}$$

is a solution of the Carathéodory–Fejér problem with the conditions

$$f^{(j)}(0) = j! d_j f(0), \quad j = 0, \dots, m,$$

where $d_0 = 1$ *and*

$$d_r = -\frac{1}{r}\sum_{j=1}^{r} d_{r-j}\gamma_j, \quad r = 1, \dots, m.$$

We use Corollary 3.3 to construct an optimal method of recovery of the functional (3.2). Set

$$W(z) := \prod_{j=1}^{n}\left(\frac{z - z_j}{1 - \overline{z}_j z}\right)^{\nu_j}, \quad W_1(z) := \left(\frac{z - \xi}{1 - \overline{\xi} z}\right)^{\nu} W(z), \quad y(z) := \frac{W'(z)}{W(z)},$$

$$\gamma_s := (-1)^s \left[\left(m + \nu - 1 + \frac{2}{p} \right) \overline{\xi}^s - \sum_{r=1}^{s} (-1)^r \binom{s}{r} \overline{\xi}^{s-r} \right.$$
$$\left. \times (1 - |\xi|^2)^r \left(\frac{y^{(r-1)}(\xi)}{(r-1)!} + \rho_r \right) \right], \quad s = 1, \dots, m, \quad (3.14)$$

where $\rho_1, \dots, \rho_m$ are defined by (3.6) for

$$d_j = \frac{\lambda_{\nu+m-j}}{\lambda_{\nu+m}}, \quad j = 1, \dots, m.$$

Substituting p for $1 - p^{-1}$ in Corollary 3.3, we obtain the existence of $b_1, \dots, b_m$ for which

$$\sum_{j=1}^{k} (\overline{b_j}^{-s} - b_j^s) + \frac{2(p-1)}{p} \sum_{j=1}^{m} b_j^s = \gamma_s, \quad s = 1, \dots, m. \quad (3.15)$$

If $p = 1$ and $k < m$, we define $b_{k+1}, \dots, b_m$ as arbitrary numbers such that $|b_j| = 1$, $j = k+1, \dots, m$.

Since for $|b_j| = 1$, $\overline{b_j}^{-s} - b_j = 0$, we can always assume that

$$\begin{aligned} &|b_j| \le 1, \quad j = 1, \dots, k, \\ &|b_j| < 1, \quad j = k+1, \dots, m. \end{aligned} \quad (3.16)$$

Thus if $p = 1$, then $k = m$.

For such $b_1, \dots, b_m$ set

$$\alpha_j := \frac{\overline{b}_j + \xi}{1 + \overline{\xi} \overline{b}_j}, \; j = 1, \dots, m, \quad \sigma := (m+1)\frac{p-2}{p} - \nu,$$
$$\Psi(z) := \prod_{j=1}^{k} \frac{z - \alpha_j}{1 - \overline{\alpha}_j z} \prod_{j=1}^{m} (1 - \overline{\alpha}_j z)^{2(p-1)/p},$$
$$C(\xi) := \frac{\lambda_{\nu+m} W(\xi)(1 - |\xi|^2)^{\sigma}}{\Psi(\xi)}$$

Theorem 3.4. *For all $1 \le p \le \infty$ the method*

$$L_\xi^\lambda f \approx \sum_{r=0}^{\nu-1} c_r(\xi) f^{(r)}(\xi) + \sum_{j=1}^{n} \sum_{r=0}^{\nu_j - 1} c_{jr}(\xi) f^{(r)}(z_j),$$

where

$$c_r(\xi) = -\frac{C(\xi)}{r!(\nu+m-r)!}\left[\frac{\Psi(z)}{W(z)(1-\bar{\xi}z)^\sigma}\right]^{(\nu+m-r)}_{|z=\xi}, \tag{3.17}$$

$$c_{jr}(\xi) = -\frac{C(\xi)}{r!(\nu_j-r-1)!}\left[\frac{\Psi(z)(1-\bar{z}_jz)^{\nu_j}}{\omega_j(z)(1-\bar{\xi}z)^\sigma(z-\xi)^{\nu+m+1}}\right]^{(\nu_j-r-1)}_{|z=z_j},$$

$$\omega_j(z) = \prod_{\substack{r=1\\ r\neq j}}^{n}\left(\frac{z-z_r}{1-\bar{z}_rz}\right)^{\nu_r},$$

is optimal for the class H_p. Moreover,

$$e(\xi,\lambda,I,H_p) = |C(\xi)|\left(\frac{1}{m!}\left[\frac{\prod_{j=1}^m(1-\bar{\alpha}_jz)(z-\alpha_j)}{(1-\bar{\xi}z)^{m+1}}\right]^{(m)}_{|z=\xi}\right)^{(p-1)/p}.$$

Proof. Set

$$g(z) := e^{i\arg C(\xi)}W_1(z)(1-\bar{\xi}z)^{-2(m+1)/p}\prod_{j=k+1}^{m}\frac{z-\alpha_j}{1-\bar{\alpha}_jz}\prod_{j=1}^{m}(1-\bar{\alpha}_jz)^{2/p}.$$

Let $1\le p<\infty$ and f is an arbitrary function from $\mathcal{H}_p$. Put

$$Jf := |C(\xi)|\frac{1}{2\pi}\int_0^{2\pi}\overline{g(e^{i\theta})}|g(e^{i\theta})|^{p-2}f(e^{i\theta})\,d\theta.$$

Using the residue theorem, we have

$$\begin{aligned}Jf &= C(\xi)\frac{1}{2\pi i}\int_{|z|=1}\frac{\Psi(z)f(z)\,dz}{W(z)(1-\bar{\xi}z)^\sigma(z-\xi)^{\nu+m+1}}\\ &= -\sum_{r=0}^{\nu+m}c_r(\xi)f^{(r)}(\xi)-\sum_{j=1}^{n}\sum_{r=0}^{\nu_j-1}c_{jr}(\xi)f^{(r)}(z_j),\end{aligned}$$

where $c_r(\xi)$ are defined by (3.17) for all $r=0,\dots,\nu+m$. We prove that

$$c_r(\xi) = -\frac{\lambda_r}{r!},\quad r=\nu,\dots,\nu+m.$$

Set $h(z) := \Psi(z)W^{-1}(z)(1-\overline{\xi}z)^{-\sigma}$ and

$$x(z) := \frac{h'(z)}{h(z)} = \sum_{j=1}^{k}\left(\frac{1}{z-\alpha_j} + \frac{\overline{\alpha}_j}{1-\overline{\alpha}_j z}\right) - \frac{2(p-1)}{p}\sum_{j=1}^{m}\frac{\overline{\alpha}_j}{1-\overline{\alpha}_j z} - y(z) + \sigma\frac{\overline{\xi}}{1-\overline{\xi}z}.$$

We have

$$\frac{x^{(r-1)}(\xi)}{(r-1)!} = \sum_{j=1}^{k}\left(\frac{(-1)^{(r-1)}}{(\xi-\alpha_j)^r} + \frac{\overline{\alpha}_j^r}{(1-\overline{\alpha}_j\xi)^r}\right) - \frac{2(p-1)}{p}\sum_{j=1}^{m}\frac{\overline{\alpha}_j^r}{(1-\overline{\alpha}_j\xi)^r} - \frac{y^{(r-1)}(\xi)}{(r-1)!} + \sigma\frac{\overline{\xi}^r}{(1-|\xi|^2)^r}.$$

Substituting α_j for $(\overline{b}_j+\xi)(1+\overline{\xi}\overline{b}_j)^{-1}$ in the last equality, we obtain

$$(1-|\xi|^2)^r\frac{x^{(r-1)}(\xi)}{(r-1)!} = -\omega_r - \frac{y^{(r-1)}(\xi)}{(r-1)!}(1-|\xi|^2)^r + \sigma\overline{\xi}^r,$$

where

$$\omega_r := \sum_{j=1}^{k}\left[(\overline{b}_j^{-1}+\overline{\xi})^r - (b_j+\overline{\xi})^r\right] + \frac{2(p-1)}{p}\sum_{j=1}^{m}(b_j+\overline{\xi})^r.$$

Analogously to (3.13) we have

$$\sum_{j=1}^{k}(\overline{b}_j^{-s} - b_j^s) + \frac{2(p-1)}{p}\sum_{j=1}^{m}b_j^s = \sum_{r=1}^{s}\binom{s}{r}(-\overline{\xi})^{s-r}\omega_r + m\frac{2(p-1)}{p}(-\overline{\xi})^s.$$

Taking into account (3.15), we obtain the following system

$$\sum_{r=1}^{s}\binom{s}{r}(-\overline{\xi})^{s-r}\omega_r = \gamma_s - m\frac{2(p-1)}{p}(-\overline{\xi})^s, \quad s = 1,\dots,m. \tag{3.18}$$

It is easily checked that this system has the unique solution

$$\omega_r = \sigma\overline{\xi}^r - (1-|\xi|^2)^r\left(\frac{y^{(r-1)}(\xi)}{(r-1)!} + \rho_r\right)$$

(it is sufficient to substitute these values into (3.18)). Thus

$$\frac{x^{(r-1)}(\xi)}{(r-1)!} = \rho_r, \quad r = 1, \ldots, m.$$

Consequently,

$$h^{(r+1)}(\xi) = (x(z)h(z))^{(r)}_{|z=\xi} = \sum_{j=0}^{r} \binom{r}{j} x^{(j)}(\xi) h^{(r-j)}(\xi)$$
$$= \sum_{j=0}^{r} \frac{r!}{(r-j)!} \rho_{j+1} h^{(r-j)}(\xi), \quad r = 0, \ldots, m-1.$$

Since

$$A \begin{pmatrix} \rho_1 \\ \rho_2 \\ \vdots \\ \rho_m \end{pmatrix} = \begin{pmatrix} d_1 \\ d_2 \\ \vdots \\ d_m \end{pmatrix},$$

we have

$$\frac{h^{(r)}(\xi)}{r!} = d_r h(\xi) = \frac{\lambda_{\nu+m-r}}{\lambda_{\nu+m}} h(\xi).$$

Thus

$$c_r(\xi) = -\frac{C(\xi)}{r!(\nu+m-r)!} h^{\nu+m-r)}(\xi) = -\frac{C(\xi)}{r!} \frac{\lambda_r}{\lambda_{\nu+m}} h(\xi) = -\frac{\lambda_r}{r!}.$$

So we proved that

$$L_\xi^\lambda f - \sum_{r=0}^{\nu-1} c_r(\xi) f^{(r)}(\xi) + \sum_{j=1}^{n} \sum_{r=0}^{\nu_j-1} c_{jr}(\xi) f^{(r)}(z_j)$$
$$= |C(\xi)| \frac{1}{2\pi} \int_0^{2\pi} \overline{g(e^{i\theta})} g(e^{i\theta})|^{p-2} f(e^{i\theta})\, d\theta.$$

To use Theorem 1.10 it remains to find $\|g\|_{\mathcal{H}_p}$. We have

$$\|g\|_{\mathcal{H}_p}^p = \frac{1}{2\pi} \int_0^{2\pi} \frac{\prod_{j=1}^m |1 - \overline{\alpha}_j e^{i\theta}|^2}{|1 - \overline{\xi} e^{i\theta}|^{2(m+1)}}\, d\theta$$
$$= \frac{1}{2\pi i} \int_{|z|=1} \frac{\prod_{j=1}^m (1 - \overline{\alpha}_j z)(z - \alpha_j)}{(1 - \overline{\xi} z)^{m+1} (z - \xi)^{m+1}}\, dz$$
$$= \frac{1}{m!} \left[\frac{\prod_{j=1}^m (1 - \overline{\alpha}_j z)(z - \alpha_j)}{(1 - \overline{\xi} z)^{m+1}} \right]^{(m)}_{|z=\xi}.$$

For $p = \infty$ we set

$$\varphi(z) := |C(\xi)| \frac{z \prod_{j=1}^{m} (z - \alpha_j)(1 - \overline{\alpha}_j z)}{(z - \xi)^{m+1} (1 - \overline{\xi} z)^{m+1}}.$$

In view of the fact that for all $|\alpha| \leq 1$ and all $|z| = 1$

$$\frac{(z - \alpha)(1 - \overline{\alpha} z)}{z} \geq 0$$

we obtain that $\varphi(e^{i\theta}) \geq 0$. Similarly to the case $1 \leq p < \infty$ we can verify that

$$\frac{1}{2\pi} \int_0^{2\pi} \overline{g(e^{i\theta})} \varphi(e^{i\theta}) f(e^{i\theta})\, d\theta$$

$$= L_\xi^\lambda f - \sum_{r=0}^{\nu-1} c_r(\xi) f^{(r)}(\xi) + \sum_{j=1}^{n} \sum_{r=0}^{\nu_j - 1} c_{jr}(\xi) f^{(r)}(z_j)$$

$$\|g\|_{\mathcal{H}_1} = |C(\xi)| \frac{1}{m!} \left[\frac{\prod_{j=1}^{m} (1 - \overline{\alpha}_j z)(z - \alpha_j)}{(1 - \overline{\xi} z)^{m+1}} \right]^{(m)}_{\Big| z = \xi}.$$

Now the assertion of the theorem follows from Theorem 1.10. □

Note that for $p = 1$ the considered problem of optimal recovery has an effective solution since such solution has the classical Carathéodory–Fejér problem (see, for example, Goluzin [1966]).

In a general case to construct an optimal method of recovery, we first solve the system (3.15) for $k = 0, 1, \ldots, m$ and then find the solution which satisfies the condition (3.16). In this connection the extremal function $g/\|g\|_{\mathcal{H}_p}$ has $m - k$ zeros in D in addition to the zeros of W_1 which prescribe by the information operator. For the fixed information operator the number of additional zeros depends on the location of ξ.

As it was mentioned, for $p = 1$, $k = m$. It means that in this case the extremal function has no additional zeros. For $1 < p \leq \infty$ the unit disk is divided into $m + 1$ sets $D_0, D_1, \ldots, D_m$ (some of them can be empty) such that if $\xi \in D_j$, then $k = m - j$. In other words, if $\xi \in D_j$, then the extremal function has exactly j additional zeros counting multiplicities. In the previous section we found D_0 and D_1 for the problem of optimal recovery of $f'(\xi)$ using the information operator

$$If = \{ f(0), f'(0), \ldots, f^{(n-1)}(0) \}.$$

Solving the extremal problem

$$\sup_{\substack{f\in H_\infty \\ f(0)=\dots=f^{(m-1)}(0)=0}} |f^{(m)}(\xi)|, \tag{3.19}$$

Goluzin [1966] found that the function $g_0(z) = z^m$ is an extremal function in this problem if and only if $|\xi| < 2^{\frac{1}{m+1}} - 1$. In our notation he proved that

$$D_0 = \{ z \in D : |z| \le 2^{\frac{1}{m+1}} - 1 \}.$$

Now we give a description of D_m in a general case. For this set the system (3.15) has the form

$$\sum_{j=1}^{m} b_j^s = \frac{p}{2(p-1)}\gamma_s, \quad s = 1, \dots, m.$$

Taking into account Corollary 3.3, we have

Theorem 3.5. *Let $1 < p \le \infty$. An extremal function of problem (3.3) has m additional zeros in D if and only if all zeros of*

$$P_m(z) = \sum_{r=0}^{m} a_r z^{m-r},$$

where $a_0 = 1$,

$$a_r = -\frac{1}{r}\frac{p}{2(p-1)}\sum_{s=1}^{r} a_{r-s}\gamma_s, \quad r = 1, \dots, m,$$

and γ_s are defined by (3.14), lie in D. Moreover, if $b_1, \dots, b_m$ are the zeros of $P_m(z)$, then

$$\alpha_j = \frac{\overline{b}_j + \xi}{1 + \overline{\xi}\overline{b}_j}, \quad j = 1, \dots, m,$$

are the additional zeros.

In particular, we can find the other sets in the problem (3.19) for $m = 2$

$$D_1 = \{ z \in D : r_0 < |z| \le r_1 \} \quad \text{and} \quad D_2 = \{ z \in D : |z| > r_1 \},$$

where $r_0 = \sqrt[3]{2} - 1 = 0.2599\dots$ and

$$r_1 = \frac{\sqrt[3]{4}}{3}\left(\sqrt[3]{\sqrt{109} + \frac{281}{27}} - \sqrt[3]{\sqrt{109} - \frac{281}{27}}\right) - \frac{4}{9} = 0.8423\dots$$

is the unique real zero of the polynomial $3t^3 + 4t^2 + 4t - 8 = 0$.

3.3 The Periodic Case for the First Derivative

In this section we prove the analogue of the result of Section 3.1 for the class $H_p^\beta(T)$. We consider the problem of optimal recovery of $f'(\xi)$, $f \in H_p^\beta(T)$, at some point $\xi \in D$ from the information operator

$$If = (f(x_1), \dots, f(x_n)),$$

where x_j are distinct points from T (for simplicity we deal with the case when all multiplicities of nodes equal 1). Denote the intrinsic error of this problem by

$$e'(\xi, H_p^\beta(T), I) := \inf_{\varphi: C^n \to C} \sup_{f \in H_p^\beta(T)} |f'(\xi) - \varphi(If)|.$$

Let K be the complete elliptic integrals of the first kind with moduli k where $k = \kappa(\beta)$ and the function κ is defined by (2.5). Set

$$W(z) := k^{n/2} \prod_{j=1}^{n} \operatorname{sn}\left(\frac{K}{\pi}(z - x_j)\right), \quad \omega_j(z) := \prod_{\substack{s=1 \\ s \neq j}}^{n} \operatorname{sn}\left(\frac{K}{\pi}(z - x_s)\right).$$

Assume that $\xi \notin \{x_1, \dots, x_n\}$ and consider the equation

$$\operatorname{sn}\gamma\left(\frac{1}{\operatorname{sn}(\gamma + K)} + k^2 \frac{p-2}{p} \operatorname{sn}(\gamma + K)\right) = \frac{\pi}{K} \frac{W'(\xi)}{W(\xi)}. \tag{3.20}$$

Denote the function in the left hand side of (3.20) by $s(\gamma)$. Since $s(\gamma)$ is a continuous function in $(-K, K)$ and $s(\gamma) \to \pm\infty$ as $\gamma \to \pm K$ there exists a solution of (3.20) $\gamma_0 \in (-K, K)$. For $\xi \in \{x_1, \dots, x_n\}$ put $\gamma_0 = K$.

Set $x_0 := \xi - \pi\gamma_0/K$,

$$w(z) := k \operatorname{sn}\left(\frac{K}{\pi}(z - \xi)\right) \operatorname{sn}\left(\frac{K}{\pi}(z - \xi + \pi)\right),$$

$$T_1 := \left\{\zeta \in T : \frac{\pi}{2kK}|W'(\zeta)| < \frac{p-1}{p}|W(\zeta)|\right\}, \qquad T_0 := T \setminus T_1,$$

$$b := \begin{cases} \dfrac{p}{p-1} \dfrac{\pi}{2kK} \dfrac{W'(\xi)}{W(\xi)}, & \xi \in T_1,\ n = 2m, \\[2ex] \dfrac{W(\xi) \operatorname{sign} W'(\xi)}{\frac{\pi}{2kK}|W'(\xi)| + \sqrt{\left(\frac{\pi}{2kK} W'(\xi)\right)^2 - \frac{p-2}{p} W^2(\xi)}}, & \xi \in T_0,\ n = 2m, \\[2ex] -w(x_0), & n = 2m - 1, \end{cases}$$

$$u_\xi(z) := \begin{cases} 1, & \xi \in T_1, \ n = 2m, \\ \dfrac{w(z)+b}{1+bw(z)}, & \xi \in T_0, \ n = 2m, \\ \dfrac{w(z)+b}{1+bw(z)} \left(\sqrt{k}\,\mathrm{sn}\left(\dfrac{K}{\pi}(z-x_0)\right)\right)^{-1}, & n = 2m-1. \end{cases}$$

Theorem 3.6. *For all* $1 \le p \le \infty$ *the method*

$$f'(\xi) \approx \sum_{j=1}^{n} c_j(\xi) f(x_j),$$

where for $\xi \ne x_j$

$$c_j(\xi) = -\frac{\pi\alpha(\xi)}{2k^{n/2+1}K} \frac{u_\xi(x_j)(1+bw(x_j))^{\frac{2(p-1)}{p}} \,\mathrm{dn}^{\frac{2(p-1)}{p}}\left(\frac{K}{\pi}(\xi - x_j)\right)}{\omega_j(x_j)\,\mathrm{sn}^2\left(\frac{K}{\pi}(\xi - x_j)\right)},$$

$$c_j(x_j) = \frac{\omega_j'(x_j)}{\omega_j(x_j)},$$

and

$$\alpha(\xi) = \begin{cases} \dfrac{2kK^2}{\pi^2}\dfrac{W(\xi)}{u_\xi(\xi)}, & \xi \notin \{x_1, \dots, x_n\}, \\ \dfrac{2k^{[\frac{n+1}{2}]}K^2}{\pi^2}\omega_j(x_j), & \xi = x_j, \ j = 1, \dots, n, \end{cases}$$

is an optimal method of recovery on the class $H_p^\beta(T)$. *Moreover, the following equality*

$$e'(\xi, H_p^\beta(T), I) = \begin{cases} \dfrac{k}{2}\left(\dfrac{2K}{\pi}\right)^{\frac{p+1}{p}} \dfrac{|W(\xi)|}{|u_\xi(\xi)|}(1+b^2)^{\frac{p-1}{p}}, & \xi \notin \{x_1, \dots, x_n\}, \\ \dfrac{k^{[\frac{n+1}{2}]}}{2}\left(\dfrac{2K}{\pi}\right)^{\frac{p+1}{p}} |\omega_j(x_j)|, & \xi = x_j \end{cases}$$

holds (here we denote by $[x]$ *the integral part of* x*).*

Proof. The function

$$v(z) := \sqrt{k}\,\mathrm{sn}\left(\frac{K}{\pi}z\right)$$

is analytic in S_β. Moreover, $v(z+2\pi) = -v(z)$ and $|v(x+i\beta)| \equiv 1$ for all $x \in R$ (see Appendix). Thus $\overline{W(z)} = W^{-1}(z)$ for $z \in \partial S_\beta$. Using the definition of b, it can be shown that $b \in [-1,1]$. Consider the function

$$g(z) := \frac{w(z)+b}{1+bw(z)} \frac{W(z)}{u_\xi(z)} (1+bw(z))^{2/p} \operatorname{dn}^{2/p}\left(\frac{K}{\pi}(z-\xi)\right).$$

Since $\operatorname{dn}(Kz/\pi)$ and $w(z)$ are 2π-periodic, $|w(z)| < 1$, $z \in S_\beta$, and $\operatorname{dn}(Kz/\pi)$ does not vanish in S_β, $g \in \mathcal{H}_p^\beta(T)$.

For $f \in H_p^\beta(T)$ and $1 \le p < \infty$ set

$$Jf := \frac{\alpha(\xi)}{4\pi} \int_{\Gamma_0} \overline{g(z)} |g(z)|^{p-2} f(z)\, dz,$$

where $\Gamma_0 := [-i\beta, 2\pi - i\beta] \cup [i\beta, 2\pi + i\beta]$. The element of integration in Jf is a 2π-periodic function. Using (2.6), we can rewrite Jf in the form

$$\begin{aligned} Jf :=& \frac{\alpha(\xi)}{4\pi i} \int_{\Gamma_\varepsilon} \frac{u_\xi(z)(1+bw(z))^{\frac{2(p-1)}{p}}}{W(z)w(z)} \\ &\times \operatorname{dn}^{\frac{p-2}{p}}\left(\frac{K}{\pi}(z-\xi)\right) \frac{\operatorname{cn}\left(\frac{K}{\pi}(z-\xi)\right)}{\operatorname{sn}\left(\frac{K}{\pi}(z-\xi)\right)} f(z)\, dz \\ =& \frac{\alpha(\xi)}{4\pi i} \int_{\Gamma_\varepsilon} \frac{u_\xi(z)(1+bw(z))^{\frac{2(p-1)}{p}} \operatorname{dn}^{\frac{2(p-1)}{p}}\left(\frac{K}{\pi}(z-\xi)\right)}{kW(z)\operatorname{sn}^2\left(\frac{K}{\pi}(z-\xi)\right)} f(z)\, dz, \end{aligned} \tag{3.21}$$

where Γ_ε is the boundary of the rectangle $-\varepsilon < \operatorname{Re} z < 2\pi - \varepsilon$, $|\operatorname{Im} z| < \beta$, and ε is such that $\xi, x_1, \dots, x_{2n}$ lie inside this rectangle. Assume that $\xi \notin \{x_1, \dots, x_{2n}\}$. By the residue theorem

$$Jf = f'(\xi) + Cf(\xi) - \sum_{j=1}^{2n} c_j(\xi) f(x_j),$$

where

$$C = \frac{\alpha(\xi)}{2k} \lim_{z\to\xi} \left(\frac{(z-\xi)^2 u_\xi(z)(1+bw(z))^{\frac{2(p-1)}{p}} \operatorname{dn}^{\frac{2(p-1)}{p}} \left(\frac{K}{\pi}(z-\xi)\right)}{W(z)\operatorname{sn}^2\left(\frac{K}{\pi}(z-\xi)\right)} \right)'$$

$$= \frac{W(\xi)}{u_\xi(\xi)} \left(\frac{u_\xi(z)(1+bw(z))^{\frac{2(p-1)}{p}}}{W(z)} \right)'\Bigg|_{z=\xi}.$$

It is not hard to check that b is defined from the condition $C = 0$. Thus we have

$$Jf = f'(\xi) - \sum_{j=1}^{n} c_j(\xi) f(x_j). \tag{3.22}$$

To calculate $\|g\|_{\mathcal{H}_p^\beta(T)}$, substitute $f(z) = g(z)$ in (3.21)

$$\alpha(\xi)\|g\|^p_{\mathcal{H}_p^\beta(T)} = \frac{\alpha(\xi)}{4\pi i} \int_{\Gamma_\varepsilon} \frac{(w(z)+b)(1+bw(z))\operatorname{dn}^2\left(\frac{K}{\pi}(z-\xi)\right)}{k\operatorname{sn}^2\left(\frac{K}{\pi}(z-\xi)\right)} dz$$

$$= \alpha(\xi)\frac{\pi}{2K}(1+b^2) \tag{3.23}$$

(we omit here some technical details concerned with the application of the residue theorem). Consequently,

$$\|g\|_{\mathcal{H}_p^\beta(T)} = \left(\frac{\pi}{2K}(1+b^2)\right)^{1/p}.$$

It remains to use Theorem 1.10.

If $\xi = x_j$, then $b = 0$ and

$$g(z) = \frac{w(z)W(z)}{u_\xi(z)} \operatorname{dn}^{2/p}\left(\frac{K}{\pi}(z-\xi)\right).$$

In this case the assertion of the theorem can be obtained by the same scheme.

For $p = \infty$ consider the integral

$$Jf := \frac{\alpha(\xi)}{4\pi} \int_{\Gamma_0} \overline{g(z)}|\varphi(z)| f(z)\, dz,$$

where

$$\varphi(z) = (1 + bw(z))^2 \operatorname{dn}^2\left(\frac{K}{\pi}(z-\xi)\right).$$

The representation (3.22) follows from (2.6) and the residue theorem. Similarly to (3.23) we obtain

$$\|\varphi\|_{\mathcal{H}_1^\beta(T)} = \frac{\pi}{2K}(1+b^2).$$

The assertion of the theorem now follows from Theorem 1.10. □

Let us consider our problem in the case when $\xi = 0$ and

$$If = I_h f := (f(-h), f(h)), \quad h \in (0, \pi).$$

In other words, we wish to construct an optimal formula of numerical differentiation at the point $\xi = 0$, using the information about function values at the points $\pm h$.

In this particular case we have

$$W(z) = k \operatorname{sn}\left(\frac{K}{\pi}(z+h)\right) \operatorname{sn}\left(\frac{K}{\pi}(z-h)\right),$$

$$W(0) = -k \operatorname{sn}^2 \frac{K}{\pi} h, \quad W'(0) = 0.$$

Moreover, $0 \in T_1$ and $b = 0$. Thus we obtain that an optimal method has the form

$$f'(0) \approx \frac{K}{\pi} \frac{f(h) - f(-h)}{\operatorname{sn} \dfrac{2K}{\pi} h} \operatorname{dn}^{\frac{2(p-1)}{p}} \frac{K}{\pi} h$$

and

$$e'(0, H_p^\beta(T), I_h) = \frac{k^2}{2}\left(\frac{2K}{\pi}\right)^{\frac{p+1}{p}} \operatorname{sn}^2 \frac{K}{\pi} h = k^2 2^{1/p}\left(\frac{K}{\pi}\right)^{\frac{3p+1}{p}} h^2 + O(h^4).$$

For optimal recovery of periodic functions the most natural system of points is an equidistant system. We will estimate the error of optimal recovery of the derivative from the information

$$If = I^{(2n)} f := (f(t_1^0), \ldots, f(t_{2n}^0)),$$

where

$$t_j^0 = (j-1)\frac{\pi}{n}, \quad j = 1, \ldots, 2n.$$

Set

$$e'_{2n}(H_p^\beta(T)) := \sup_{\xi \in T} e'(\xi, H_p^\beta(T), I^{(2n)}).$$

Theorem 3.7. *For all $\beta > 0$*

$$e'_{2n}(H^{\beta}_{\infty}(T)) = \sqrt{\lambda}\frac{2n\Lambda}{\pi} = 2ne^{-\beta n} + O(ne^{-5\beta n}),$$

$$e'_{2n}(H^{\beta}_{2}(T)) = \sqrt{\frac{2K\lambda}{\pi}\frac{2n\Lambda}{\pi}} = \sqrt{\frac{2K}{\pi}}2ne^{-\beta n} + O(ne^{-5\beta n}),$$

where Λ is the complete elliptic integral of the first kind for modulus

$$\lambda = \kappa(2\beta n) = 4e^{-2\beta n}\left(\frac{\sum_{m=0}^{\infty} e^{-4\beta nm(m+1)}}{1+2\sum_{m=1}^{\infty} e^{-4\beta nm^2}}\right)^2.$$

Proof. Using the first principal transform of elliptic functions of degree $2n$ (see Appendix), we find

$$\begin{aligned} W\left(z - \frac{\pi}{2n}\right) &= k^n \prod_{j=1}^{2n} \operatorname{sn}\left(\frac{K}{\pi}z - \frac{2j-1}{2n}K\right) \\ &= (-1)^n k^n \prod_{j=1}^{n} \operatorname{sn}\left(\frac{K}{\pi}z - \frac{2j-1}{2n}K\right)\operatorname{sn}\left(\frac{K}{\pi}z + \frac{2j-1}{2n}K\right) \\ &= k^n \prod_{j=1}^{n} \frac{\operatorname{sn}^2\frac{2j-1}{2n}K - \operatorname{sn}^2\frac{K}{\pi}z}{1 - k^2\operatorname{sn}^2\frac{2j-1}{2n}K\operatorname{sn}^2\frac{K}{\pi}z} = \sqrt{\lambda}\operatorname{sn}\left(\frac{2n\Lambda}{\pi}z + \Lambda, \lambda\right). \end{aligned}$$

Hence

$$W(z) = -\sqrt{\lambda}\operatorname{sn}\left(\frac{2n\Lambda}{\pi}z, \lambda\right).$$

In view of the equalities

$$\frac{d}{dt}\operatorname{sn}(t,\lambda) = \operatorname{cn}(t,\lambda)\operatorname{dn}(t,\lambda) = \sqrt{(1-\operatorname{sn}^2(t,\lambda))(1-\lambda^2\operatorname{sn}^2(t,\lambda))},$$

from Theorem 3.6 we obtain

$$e'_{2n}(H^{\beta}_{p}(T)) = \sup_{s\in[0,1]} \frac{k}{2}\left(\frac{2K}{\pi}\right)^{\frac{1+p}{p}} \sqrt{\lambda}\Phi_p(s),$$

where

$$\Phi_p(s)=\begin{cases} s\left(1+\left(\dfrac{pa}{p-1}\right)^2\dfrac{(1-s^2)(1-\lambda^2s^2)}{s^2}\right)^{\frac{p-1}{p}}, & s\in S_p,\\ \gamma(s)\left(1+\dfrac{s^2}{\gamma^2(s)}\right)^{\frac{p-1}{p}}, & s\in[0,1]\setminus S_p,\end{cases}$$

$$a=\frac{n\Lambda}{kK},\quad S_p=\left\{s\in[0,1]:a^2(1-s^2)(1-\lambda^2s^2)<\left(\frac{p-1}{p}\right)^2s^2\right\},$$

$$\gamma(s)=a\sqrt{(1-s^2)(1-\lambda^2s^2)}+\sqrt{a^2(1-s^2)(1-\lambda^2s^2)-\frac{p-2}{p}s^2}.$$

Let us begin with the case $p=2$. It is easy to check that

$$\Phi_2(s)=\sqrt{s^2+4a^2(1-s^2)(1-\lambda^2s^2)}.$$

From properties of the first principal transform of elliptic functions of degree $2n$ it follows that

$$\frac{2n\Lambda}{K}=\prod_{j=1}^{n}\frac{\operatorname{sn}^2\frac{j}{n}K}{\operatorname{sn}^2\frac{2j-1}{2n}K}>1.$$

Hence $2a>1$ and

$$\Phi_2^2(s)\le s^2+4a^2(1-s^2)\le 4a^2.$$

This estimate is attained for $s=0$. Thus

$$e'_{2n}(H_2^\beta(T))=\sqrt{\frac{2K\lambda}{\pi}\frac{2n\Lambda}{\pi}}.$$

The asymptotic equality follows from the fact that

$$\sqrt{\lambda}=2e^{-\beta n}+O\left(e^{-5\beta n}\right),$$
$$\Lambda=\frac{\pi}{2}+O\left(e^{-4\beta n}\right).$$

Let $p=\infty$. It can be easily shown that $S_\infty=(s^*,1]$ where s^* is the unique solution of the equation

$$a^2(1-s^2)(1-\lambda^2s^2)=s^2.$$

We have

$$\Phi_\infty(s) = \begin{cases} 2a\sqrt{(1-s^2)(1-\lambda^2 s^2)}, & s \in [0, s^*], \\ s + a^2 \dfrac{(1-s^2)(1-\lambda^2 s^2)}{s}, & s \in (s^*, 1]. \end{cases}$$

Since the function

$$F(s) := s + a^2 \frac{(1-s^2)(1-\lambda^2 s^2)}{s}$$

is convex for $s \in (0,1)$, we obtain

$$\max_{s \in [s^*, 1]} F(s) = \max\{F(s^*), F(1)\} = \max\{\Phi_\infty(s^*), 1\}.$$

The function $\Phi_\infty(s)$ decreases while $s \in [0, s^*]$. Consequently,

$$\max_{s \in [0,1]} \Phi_\infty(s) = \max\{\Phi_\infty(0), 1\} = 2a.$$

□

3.4 Noisy Information

Consider now the problem of optimal recovery of derivatives of functions by their values given with an error in the uniform norm. Let W be a convex balanced set of sufficiently smooth functions defined on the domain $\Omega \in C$ containing some set E. For $\xi \in \Omega$ set

$$e^{(k)}(\xi, E, W, \delta) := \inf_{\varphi:\, C(E) \to C} \sup_{f \in W} \sup_{\substack{\tilde{f} \in C(E) \\ \|f - \tilde{f}\|_{C(E)} \le \delta}} |f^{(k)}(\xi) - \varphi(\tilde{f})|. \tag{3.24}$$

Thus we consider the problem of optimal recovery of the kth order derivative of the function $f \in W$ knowing the function $\tilde{f} \in C(E)$ such that

$$\sup_{\xi \in E} |f(\xi) - \tilde{f}(\xi)| \le \delta.$$

Let $W = H_\infty$ and $E = (-1, 1)$. Then in view of duality (see Theorem 1.6) we have

$$e^{(k)}(\xi, (-1,1), H_\infty, \delta) = \sup_{\substack{f \in H_\infty \\ \|f\|_{C(-1,1)} \le \delta}} |f^{(k)}(\xi)|.$$

Consequently, for $\delta \geq 1$

$$e^{(k)}(\xi, (-1,1), H_\infty, \delta) = \sup_{f \in H_\infty} |f^{(k)}(\xi)|,$$

and the method $\varphi(\tilde{f}) \equiv 0$ is an optimal method. The solution of this extremal problem can be obtained from Theorem 3.4.

In what follows we shall consider the case when $0 < \delta < 1$. In this case it turns out that an optimal method of recovery does not use all values of $\tilde{f}$. It uses only the values at a discrete system of points from $(-1,1)$. The density of these points increases while $\delta \downarrow 0$.

Theorem 3.8. *For all $0 < \delta < 1$ and $\xi \in (-1,1)$ the method*

$$f'(\xi) \approx \frac{2\pi}{\Lambda'(1-\delta^4)(1-\xi^2)} \sum_{j=-\infty}^{\infty} \frac{(-1)^{j+1}}{\sinh^2\left((2j-1)\dfrac{\pi\Lambda}{\Lambda'}\right)} \tilde{f}\left(\frac{b_j+\xi}{1+\xi b_j}\right),$$

where b_j are defined by (2.18) and Λ, Λ' are the complete elliptic integrals of the first kind for the moduli $\lambda = \delta^2$, $\lambda' = \sqrt{1-\delta^4}$, respectively, is an optimal method of recovery on the class H_∞ from values on the interval $(-1,1)$ given with an error δ. Moreover,

$$e^{(1)}(\xi, (-1,1), H_\infty, \delta) = \frac{2\delta\Lambda'}{\pi(1-\xi^2)} = \frac{4}{\pi(1-\xi^2)}\delta \log\frac{2}{\delta} + O\left(\delta^5 \log\frac{1}{\delta}\right).$$

Proof. Put

$$B_1(z) := \prod_{j=1}^{\infty} \frac{b_j^2 - z^2}{1 - b_j^2 z^2}, \qquad h(z) := \prod_{j=1}^{\infty} \left(\frac{1-a_j^2 z^2}{1-b_j^2 z^2}\right)^2, \tag{3.25}$$

$$\psi(z) := \frac{B_0(z) h(z)}{z B_1(z)}, \tag{3.26}$$

where a_j and b_j are defined by (2.15) and (2.18) while B_0 is defined by (2.16). In the same way as the representation (2.17) was obtained it can be shown that

$$B_1(z) = \sqrt{\lambda}\, \mathrm{sn}\left(\frac{2\Lambda'}{\pi} \operatorname{arctanh} z + \Lambda, \lambda\right), \tag{3.27}$$

and

$$h(z) = \frac{1-z^2}{\mathrm{dn}^2\left(\dfrac{2\Lambda'}{\pi} \operatorname{arctanh} z, \lambda\right)}. \tag{3.28}$$

Thus,

$$\psi(z)=\frac{1-z^2}{z}\frac{\operatorname{sn}\left(\frac{2\Lambda'}{\pi}\operatorname{arctanh} z,\lambda\right)}{\operatorname{cn}\left(\frac{2\Lambda'}{\pi}\operatorname{arctanh} z,\lambda\right)\operatorname{dn}\left(\frac{2\Lambda'}{\pi}\operatorname{arctanh} z,\lambda\right)}.$$

In view of the Gauss transform of elliptic functions (see Appendix) we get

$$\begin{aligned}\psi(z)&=\frac{1-z^2}{z}\frac{1}{1+\lambda}\frac{\operatorname{sn}\left(\frac{4L'}{\pi}\operatorname{arctanh} z,l\right)}{\operatorname{cn}\left(\frac{4L'}{\pi}\operatorname{arctanh} z,l\right)}\\&=-\frac{1-z^2}{z}\frac{i}{(1+\lambda)\operatorname{dn}\left(\frac{4L'}{\pi}\operatorname{arctanh} z+iL',l\right)},\end{aligned}$$

where $l=2\sqrt{\lambda}/(1+\lambda)$ and L, L' are the complete elliptic integrals of the first kind for the moduli l, $l'=\sqrt{1-l^2}$, respectively. If $z=e^{i\theta}$, $\theta\in(0,\pi)\cup(\pi,2\pi)$, then $\operatorname{arctanh} z=x+i\frac{\pi}{4}\operatorname{sign}\sin\theta$, where $x=\frac{1}{2}\log\left|\cot\frac{\theta}{2}\right|$. Consequently,

$$\begin{aligned}\psi(e^{i\theta})&=-\frac{2\sin\theta}{(1+\lambda)\operatorname{dn}\left(\frac{4L'}{\pi}x+iL'(1+\operatorname{sign}\sin\theta),l\right)}\\&=\frac{2|\sin\theta|}{(1+\lambda)\operatorname{dn}\left(\frac{4L'}{\pi}x,l\right)}.\end{aligned}$$

Since for all $u\in R$

$$1\geq \operatorname{dn}(u,l)\geq l'=\frac{1-\lambda}{1+\lambda},$$

for almost all $\theta\in T$, $\psi(e^{i\theta})>0$. Moreover, $\psi(e^{i\theta})\in L_1(T)$.

For $f\in H_\infty$ put

$$Jf:=\frac{\delta}{2\pi}\int_0^{2\pi}\overline{B_0(e^{i\theta})}\psi(e^{i\theta})f(e^{i\theta})\,d\theta.$$

It is easy to check that this integral can be written in the form

$$Jf=\frac{\delta}{2\pi i}\int_{|z|=1}\frac{h(z)}{B_1(z)z^2}f(z)\,dz. \tag{3.29}$$

In view of (3.27)

$$B_1(a_j) = (-1)^j \sqrt{\lambda}.$$

After applying Lemma 2.5 to the integral (3.29), we find

$$Jf = f'(0) + \delta \sum_{j=-\infty}^{\infty} \frac{h(b_j)}{B_1'(b_j) b_j^2} f(b_j).$$

From (3.27) and (3.28) we have

$$B_1'(b_j) = (-1)^j \sqrt{\lambda} \frac{2\Lambda'}{\pi(1-b_j^2)}, \qquad h(b_j) = \frac{1-b_j^2}{1-\lambda^2}. \tag{3.30}$$

Thus

$$Jf = f'(0) - \varphi_0(If),$$

where

$$\varphi_0(y) = \frac{2\pi}{\Lambda'(1-\delta^4)} \sum_{j=-\infty}^{\infty} \frac{(-1)^{j+1}}{\sinh^2\left((2j-1)\frac{\pi\Lambda}{\Lambda'}\right)} y(b_j), \qquad y \in C(-1,1).$$

Moreover,

$$\varphi_0(B_0) = \delta \|\varphi_0\|.$$

Using Theorem 1.9, we obtain that the method φ_0 is optimal for $\xi = 0$ and that $B_0(z)$ is an extremal function. Hence

$$e^{(1)}(0,(-1,1),H_\infty,\delta) = B_0'(0) = \delta \frac{2\Lambda'}{\pi}.$$

The asymptotic equality for the intrinsic error follows from the equalities (see (A.41))

$$\Lambda' = \log \frac{4}{\lambda} + O\left(\lambda^2 \log \frac{4}{\lambda}\right) = \log \frac{4}{\delta^2} + O\left(\delta^4 \log \frac{1}{\delta}\right).$$

In the case of arbitrary $\xi \in (-1,1)$ the statement of theorem can be obtained by considering the function $g(z) = f(w(z))$ where

$$w(z) := \frac{z+\xi}{1+\xi z}$$

is a conformal transformation of the unit disk D. For this function we have $g'(0) = (1-\xi^2) f'(\xi)$. □

Using the conformal map of the strip S_β onto the unit disk, from Theorem 3.8 we obtain

Corollary 3.9. *For all* $0 < \delta < 1$ *and* $t \in R$ *the method*

$$f'(t) \approx \frac{\pi^2}{2\beta\Lambda'(1-\delta^4)} \sum_{j=-\infty}^{\infty} \frac{(-1)^{j+1}}{\sinh^2\left((2j-1)\dfrac{\pi\Lambda}{\Lambda'}\right)} \tilde{f}\left(t + (2j-1)\frac{2\beta\Lambda}{\Lambda'}\right)$$

is an optimal method of recovery on the class $H_\infty^\beta(R)$ *from values on* R *given with an error* δ. *Moreover,*

$$e^{(1)}(t, R, H_\infty^\beta(R), \delta) = \frac{\delta\Lambda'}{2\beta} = \frac{1}{\beta}\delta \log\frac{2}{\delta} + O\left(\delta^5 \log\frac{1}{\delta}\right).$$

Now we consider the case $k = 2$. It turns out that, by contrast with the previous case $k = 1$, when $k = 2$ there arises the "switching" effect, which consists in the existence of $\delta_0 \in (0, 1)$ such that for $0 < \delta \le \delta_0$ and $\delta_0 < \delta < 1$ the extremal functions and the methods of recovery are qualitatively different.

We set

$$C(\delta) := \frac{8}{3}\left[\frac{1-5\delta^4}{2}\left(\frac{\Lambda'}{\pi}\right)^2 - 1\right].$$

As above we denote by Λ and Λ' the complete elliptic integrals of the first kind for the moduli $\lambda = \delta^2$ and $\lambda' = \sqrt{1-\delta^4}$, respectively. Since Λ' is monotone decreasing for $\delta \in (0, 1)$, it follows that the equation

$$\frac{1-5\delta^4}{2}\left(\frac{\Lambda'}{\pi}\right)^2 = 1 \tag{3.31}$$

has a unique solution $\delta_0 \in (0, 1)$ (computations show that $\delta_0 = 0{,}2145\ldots$). Thus $C(\delta_0) = 0$. Moreover, $C(\delta) > 0$ for $\delta \in (0, \delta_0)$ and $C(\delta) < 0$ for $\delta \in (\delta_0, 1)$.

Consider the function

$$F(x) := \frac{4}{\sinh^2 \dfrac{\pi}{\Lambda'}x}\left[1 - \frac{\Lambda'}{2\pi}\sinh\frac{2\pi}{\lambda'}x \frac{\operatorname{cn}(x,\delta^2)(1+\delta^4\operatorname{sn}^2(x,\delta^2))}{\operatorname{sn}(x,\delta^2)\operatorname{dn}(x,\delta^2)}\right].$$

Using the expansions (see (A.45))

$$\sinh x = x + \frac{x^3}{6} + O(x^5), \qquad \operatorname{sn}(x,\delta^2) = x - \frac{1+\delta^4}{6}x^3 + O(x^5),$$

$$\operatorname{cn}(x,\delta^2) = 1 - \frac{x^2}{2} + O(x^4), \qquad \operatorname{dn}(x,\delta^2) = 1 - \delta^4\frac{x^2}{2} + O(x^4),$$

we find that

$$F(x) = C(\delta) + O(x^2).$$

Thus, for $\delta \in (0, \delta_0)$ we have $F(0) < 0$. Since

$$F(\Lambda) = 4 \sinh^{-2} \frac{\pi \Lambda}{\Lambda'} > 0,$$

it follows that for any $\delta \in (\delta_0, 1)$ there exists $\gamma \in (0, \Lambda)$ for which $F(\gamma) > 0$.

Theorem 3.10. *For all $t \in R$ the methods*

$$f''(t) \approx -\frac{\pi^2}{16\beta^2} C(\delta) \tilde{f}(t) + \frac{\pi^2}{2\beta^2} \sum_{j \in Z \setminus \{0\}} \frac{(-1)^{j+1}}{\sinh^2 2j \frac{\pi \Lambda}{\Lambda'}} \tilde{f}\left(t + j\frac{4\beta\Lambda}{\Lambda'}\right)$$

for $0 < \delta \leq \delta_0$ and

$$f''(t) \approx \frac{\pi^3}{4\beta^2 \Lambda'} \frac{\cosh^3 \gamma \frac{\pi}{\Lambda'} \operatorname{sn}(\gamma, \delta^2)}{\tanh \gamma \frac{\pi}{2\Lambda'} \operatorname{dn}^2(\gamma, \delta^2)}$$

$$\times \sum_{j \in Z \setminus \{0\}} \frac{(-1)^{j+1} \sinh^2 2j \frac{\pi \Lambda}{\Lambda'}}{\sinh^2 \left((2j\Lambda - \gamma)\frac{\pi}{\Lambda'}\right) \sinh^2 \left((2j\Lambda + \gamma)\frac{\pi}{\Lambda'}\right)} \tilde{f}(t + t_j)$$

for $\delta_0 < \delta < 1$, where

$$t_j = \frac{4\beta}{\pi} \operatorname{arctanh} \left[\tanh\left((2j\Lambda - \gamma)\frac{\pi}{\Lambda'}\right) \tanh\left((2j\Lambda + \gamma)\frac{\pi}{\Lambda'}\right)\right]^{1/2} \operatorname{sign} j,$$

are optimal methods of recovery on the class $H^{\beta}_{\infty}(R)$ from values on R given with an error δ. Moreover,

$$e^{(2)}(t, R, H^{\beta}_{\infty}(R), \delta) = \begin{cases} \dfrac{\Lambda'^2}{4\beta^2} \delta(1 - \delta^4), & 0 < \delta \leq \delta_0, \\ \dfrac{\pi \Lambda'}{4\beta^2} \dfrac{\delta(1 - \delta^4) \operatorname{sn}(\gamma, \delta^2)}{\tanh \gamma \frac{\pi}{\Lambda'} \operatorname{dn}(\gamma, \delta^2)}, & \delta_0 \leq \delta < 1. \end{cases}$$

Proof. In view of translation invariance it suffices to prove the assertion of the theorem for $t = 0$. Using the conformal map

$$z = \tanh\left(\frac{\pi}{4\beta} t\right),$$

our problem of recovery can be reduced to the problem of recovering the value of

$$\frac{\pi^2}{16\beta^2} f''(0)$$

on the class H_∞ from values on the interval $(-1,1)$, given with an error δ. For $f \in H_\infty$ we set

$$Jf := \frac{\alpha}{2\pi} \int_{|z|=1} \frac{\Psi(z) f(z)}{z^3}\, dz, \tag{3.32}$$

where

$$\alpha = \begin{cases} \delta \dfrac{4\Lambda'}{\pi}, & 0 < \delta \le \delta_0, \\ \delta \dfrac{4\operatorname{sn}(\gamma, \delta^2)}{\sinh \gamma \frac{\pi}{\Lambda'} \operatorname{dn}^2(\gamma, \delta^2)}, & \delta_0 < \delta < 1, \end{cases}$$

$$\Psi(z) = (1 - z^2)^2 h_0\left(-\frac{z^2 + a^2}{1 + a^2 z^2}\right) B_2^{-1}\left(-\frac{z^2 + a^2}{1 + a^2 z^2}\right),$$

$$h_0(z) = \prod_{j=1}^{\infty} \left(\frac{1 + b_j^2 z}{1 + a_j^2 z}\right)^2, \quad B_2(z) = \prod_{j=1}^{\infty} \frac{z + a_j^2}{1 + a_j^2 z},$$

$$a = \begin{cases} 0, & 0 < \delta \le \delta_0, \\ \tanh \gamma \dfrac{\pi}{2\Lambda'}, & \delta_0 < \delta < 1 \end{cases}$$

(a_j and b_j are defined by (2.15) and (2.18)). Set

$$B_3(z) := \prod_{j=1}^{\infty} \frac{z + b_j^2}{1 + b_j^2 z},$$

Using the product expansion of elliptic functions (see Appendix), we can show that the following representations hold:

$$\begin{gathered} B_2(z) = \delta \coth v \operatorname{sn}\left(\frac{2\Lambda'}{\pi} v, \delta^2\right), \quad B_3(z) = \delta \operatorname{sn}\left(\frac{2\Lambda'}{\pi} v + \Lambda, \delta^2\right), \\ h_0(z) = \cosh^2 v \operatorname{dn}^2\left(\frac{2\Lambda'}{\pi} v\right), \quad z = -\tanh^2 v. \end{gathered} \tag{3.33}$$

Applying Lemma 2.5 to the integral (3.32), we have

$$Jf = \frac{\alpha}{2}(\Psi(0) f''(0) + 2\Psi'(0) f'(0)) - S(\delta) f,$$

where

$$S(\delta)f = -\frac{\alpha}{2}\Psi''(0)f(0) + \frac{\alpha}{2}\sum_{j\in Z\setminus\{0\}} \frac{(1-x_j^2)^2(1+a^2x_j^2)^2h_0(-a_j^2)}{x_j^4(1-a^4)B_2'(-a_j^2)}f(x_j),$$

$$x_j = \left(\frac{a_j^2-a^2}{1-a^2a_j^2}\right)^{1/2} \operatorname{sign} j.$$

It is easy to see that for $\delta_0 < \delta < 1$ we have

$$x_j = \tanh\left(\frac{\pi}{4\beta}t_j\right),$$

while for $0 < \delta \le \delta_0$ we have $x_j = a_j$.

It follows from the representation (3.33) that

$$h_0(-a_j^2) = \frac{1}{1-a_j^2}, \quad B_2'(-a_j^2) = (-1)^{j+1}\delta\frac{\Lambda'}{\pi}\frac{1}{a_j^2(1-a_j^2)},$$

$$\Psi(z) = \frac{(1-z^2)(1+a^2z^2)\operatorname{dn}^2\left(\frac{2\Lambda'}{\pi}v,\delta^2\right)\tanh v}{\delta(1-a^2)\operatorname{sn}\left(\frac{2\Lambda'}{\pi}v,\delta^2\right)}, \qquad \frac{z^2+a^2}{1+a^2z^2} = \tanh^2 v.$$

Hence $\Psi(0) = 2\alpha^{-1}$, $\Psi'(0) = 0$, $\frac{\alpha}{2}\Psi''(0) = C(\delta)$ for $0 < \delta \le \delta_0$, and $\frac{\alpha}{2}\Psi''(0) = F(\gamma) = 0$ for $\delta_0 < \delta < 1$. Thus

$$Jf = f''(0) - S(\delta)f.$$

After applying transformations for $S(\delta)f$, we obtain the following equalities:

$$S(\delta) = -C(\delta)f(0) + 8\sum_{j\in Z\setminus\{0\}} \frac{(-1)^{j+1}}{\sinh^2 2j\frac{\pi\Lambda}{\Lambda'}}f(a_j)$$

if $0 < \delta \le \delta_0$, and

$$S(\delta) = \frac{4\pi}{\Lambda'}\frac{\cosh^3\gamma\frac{\pi}{\Lambda'}\operatorname{sn}(\gamma,\delta^2)}{\tanh\gamma\frac{\pi}{2\Lambda'}\operatorname{dn}^2(\gamma,\delta^2)}$$

$$\times\sum_{j\in Z\setminus\{0\}} \frac{(-1)^{j+1}\sinh^2 2j\frac{\pi\Lambda}{\Lambda'}}{\sinh^2\left((2j\Lambda-\gamma)\frac{\pi}{\Lambda'}\right)\sinh^2\left((2j\Lambda+\gamma)\frac{\pi}{\Lambda'}\right)}f(x_j)$$

if $\delta_0 < \delta < 1$.

The integral (3.32) can be written in the form

$$Jf = \frac{\alpha}{2\pi}\int_0^{2\pi} \overline{g(e^{i\theta})}\Phi(e^{i\theta})f(e^{i\theta})\,d\theta,$$

where

$$g(z) = -B_3\left(-\frac{z^2+a^2}{1+a^2z^2}\right), \quad \Phi(z) = \frac{\Psi(z)g(z)}{z^2}.$$

We claim that $\Phi(e^{i\theta}) \in L_1[0,2\pi]$ and $\Phi(e^{i\theta}) > 0$ almost everywhere. In view of the easily verified equality

$$\Phi(z) = \frac{(z^2+a^2)(1+a^2z^2)}{(1-a^2)^2z^2}\Phi_0\left(-\frac{z^2+a^2}{1+a^2z^2}\right),$$

in which

$$\Phi_0(z) = \frac{(1+z)^2}{z}\frac{B_3(z)h_0(z)}{B_2(z)},$$

it suffices to prove that Φ_0 enjoys this property. Using the Gauss transform (see Appendix), it follows from (3.33) that

$$\Phi_0(z) = -\frac{2}{\sinh 2v}\frac{\operatorname{cn}\left(\frac{2\Lambda'}{\pi}v,\delta^2\right)\operatorname{dn}\left(\frac{2\Lambda'}{\pi}v,\delta^2\right)}{\operatorname{sn}\left(\frac{2\Lambda'}{\pi}v,\delta^2\right)}$$

$$= -\frac{2(1+\delta^2)}{\sinh 2v}\frac{\operatorname{cn}\left(\frac{4L'}{\pi}v,l\right)}{\operatorname{sn}\left(\frac{4L'}{\pi}v,l\right)} = -i\frac{2(1+\delta^2)}{\sinh 2v}\operatorname{dn}\left(\frac{4L'}{\pi}v+iL',l\right),$$

where $l = 2\delta/(1+\delta^2)$ and L, L' are the complete elliptic integrals of the first kind for the moduli l, $l' = \sqrt{1-l^2}$, respectively. Let $z = e^{i\theta}$, $\theta \in (-\pi,\pi)$. Then v can be chosen from the condition $\tanh v = e^{i(\theta+\pi)/2}$. Hence $v = x + i\pi/4$, where

$$x = \frac{1}{2}\log\left|\cot\frac{\theta+\pi}{4}\right|.$$

Consequently,

$$\Phi_0(e^{i\theta}) = \frac{2(1+\delta^2)}{\cosh 2x}\operatorname{dn}\left(\frac{4L'}{\pi}x,l\right).$$

It is clear from the last equality that for $\theta \in (-\pi, \pi)$

$$0 < \Phi_0(e^{i\theta}) \leq 2(1 + \delta^2).$$

In order to make use of Theorem 1.9 it remains to verify the equality $S(\delta)g = \delta\|S(\delta)\|$, which follows from the fact that

$$B_3(-a_j^2) = (-1)^j \delta.$$

The same theorem also yields

$$e^{(2)}(0, (-1, 1), H_\infty, \delta) = g''(0).$$

The expression for $g''(0)$ is easily found by using the representation (3.33).

Passing from the class H_∞ to the class $H_\infty^\beta(R)$, we see that the method

$$f''(0) \approx \frac{\pi^2}{16\beta^2} S(\delta)\tilde{y},$$

where

$$\tilde{y}(z) = \tilde{f}\left(\frac{4\beta}{\pi} \operatorname{arctanh} z\right),$$

is optimal, and

$$e^{(2)}(0, R, H_\infty^\beta(R), \delta) = \frac{\pi^2}{16\beta^2} e^{(2)}(0, (-1, 1), H_\infty, \delta).$$

□

3.5 Kolmogorov Inequalities

Kolmogorov [1939] posed and solved the following problem: to find

$$\sup \|f^{(k)}\|_\infty, \quad 1 \leq k \leq r - 1,$$

over all functions that have absolutely continuous derivatives up to $(r-1)$th order and satisfy the condition

$$\|f\|_\infty \leq A_0, \qquad \|f^{(r)}\|_\infty \leq A_r,$$

where $\|\cdot\|_\infty$ is the norm in $L_\infty(R)$.

Denote by $H_\infty^{r,\beta}(R)$ the class of functions analytic in the strip S_β and satisfying the condition

$$|f^{(r)}(z)| \leq 1, \quad z \in S_\beta.$$

We shall be interested in the extremal problem

$$\sup_{\substack{f\in H^{r,\beta}_\infty(R)\\ \|f\|_\infty\le\delta}} \|f^{(k)}\|_\infty .$$

This problem closely connected with the problem of optimal recovery of derivatives from incorrect information about a function. The intrinsic error of optimal recovery of $f^{(k)}(t)$ from the trace $f_{|R}$ known with an error δ in the uniform norm is defined by

$$e^{(k)}(t,R,H^{r,\beta}_\infty(R),\delta) := \inf_{\varphi:\, C(R)\to C}\ \sup_{f\in H^{r,\beta}_\infty(R)}\ \sup_{\substack{\tilde f\in C(R)\\ \|f-\tilde f\|_\infty\le\delta}} |f^{(k)}(t)-\varphi(\tilde f)|$$

(see (3.24)). For $r=0$ and $k=1,2$ the solution of this problem was obtained in Section 3.4. For $t\in R$ from duality (see Theorem 1.6) and translation invariance we have

$$e^{(k)}(t,R,H^{r,\beta}_\infty(R),\delta) = \sup_{\substack{f\in H^{r,\beta}_\infty(R)\\ \|f\|_\infty\le\delta}} |f^{(k)}(t)| = \sup_{\substack{f\in H^{r,\beta}_\infty(R)\\ \|f\|_\infty\le\delta}} \|f^{(k)}\|_\infty .$$

Let Λ and Λ' be the complete elliptic integrals of the first kind for the moduli $\lambda\in(0,1)$ and $\lambda'=\sqrt{1-\lambda^2}$, respectively. Consider the function

$$\phi_\lambda(z) := \sqrt{\lambda}\,\mathrm{sn}\left(\frac{\Lambda'}{2\beta}z,\lambda\right).$$

From the properties of elliptic functions it follows that for all $x\in R$, $|\phi_\lambda(x+i\beta)|=1$. Thus for any $\lambda\in(0,1)$, $\phi_\lambda\in H^{\beta}_\infty(R)$.

Put $\Phi_{\lambda,0,\beta}(z) := \phi_\lambda(z)$,

$$\begin{aligned}\Phi_{\lambda,2j-1,\beta}(z) &:= \int_{2\Lambda\beta/\Lambda'}^{z} \Phi_{\lambda,2j-2,\beta}(u)\,du,\\ \Phi_{\lambda,2j,\beta}(z) &:= \int_0^z \Phi_{\lambda,2j-1,\beta}(u)\,du,\end{aligned}\qquad j=1,2,\ldots .$$

The properties of $\Phi_{\lambda,r,\beta}$ are analogous to the properties of the perfect Euler splines (see Korneichuk [1987]). Using the expansion of the elliptic sine in a Fourier series (see (A.47))

$$\mathrm{sn}\left(\frac{\Lambda'}{2\beta}z,\lambda\right) = \frac{\pi}{\lambda\Lambda}\sum_{n=0}^{\infty}\frac{\sin\left((2n+1)\dfrac{\pi\Lambda'}{4\Lambda\beta}z\right)}{\sinh\left((2n+1)\dfrac{\pi\Lambda'}{2\Lambda}\right)},$$

we obtain

$$\begin{aligned}\Phi_{\lambda,r,\beta}(z) &= \frac{\pi}{\sqrt{\lambda}\Lambda}\left(\frac{4\Lambda\beta}{\pi\Lambda'}\right)^r \sum_{n=0}^{\infty} \frac{\sin\left((2n+1)\dfrac{\pi\Lambda'}{4\Lambda\beta}z - \dfrac{\pi r}{2}\right)}{(2n+1)^r \sinh\left((2n+1)\dfrac{\pi\Lambda'}{2\Lambda}\right)},\\ \|\Phi_{\lambda,r,\beta}\|_\infty &= \frac{\pi}{\sqrt{\lambda}\Lambda}\left(\frac{4\Lambda\beta}{\pi\Lambda'}\right)^r \sum_{n=0}^{\infty} \frac{(-1)^{n(r+1)}}{(2n+1)^r \sinh\left((2n+1)\dfrac{\pi\Lambda'}{2\Lambda}\right)}.\end{aligned} \tag{3.34}$$

Theorem 3.11. *Let $\delta \in (0,\infty)$ for $r \geq 1$, and let $\delta \in (0,1)$ for $r = 0$. Then for any $1 \leq k \leq r+1$*

$$\begin{aligned}\sup_{\substack{f\in H_\infty^{r,\beta}(R)\\ \|f\|_\infty\leq\delta}} \|f^{(k)}\|_\infty &= \|\Phi^{(k)}_{\lambda,r,\beta}\|_\infty\\ &= \frac{\pi}{\sqrt{\lambda}\Lambda}\left(\frac{4\Lambda\beta}{\pi\Lambda'}\right)^{r-k} \sum_{n=0}^{\infty} \frac{(-1)^{n(r-k+1)}}{(2n+1)^{r-k} \sinh\left((2n+1)\dfrac{\pi\Lambda'}{2\Lambda}\right)},\end{aligned} \tag{3.35}$$

where λ satisfies the equality

$$\|\Phi_{\lambda,r,\beta}\|_\infty = \delta. \tag{3.36}$$

In the proof of Theorem 3.11 we use the inductive scheme applied in the Kolmogorov comparison theorem (see, for example, Korneichuk [1976]). However, unlike the smooth case, in the considered case the first step of the induction is the most difficult one. First we prove a number of preliminary results.

Denote by $H_{\infty,R}$ the set of functions from H_∞ that are real on the real axis.

Proposition 3.12. *Let $f \in H_{\infty,R}$, and let*

$$\|f\|_{C(-1,1)} \leq \|B_0\|_{C(-1,1)}, \tag{3.37}$$

for some $\lambda \in (0,1)$, where

$$\|x\|_{C(-1,1)} := \sup_{t\in(-1,1)} |x(t)|$$

and B_0 is defined by (2.16). If $\xi \in (-1,1)$ and $f(\xi) = B_0(\xi)$, then

$$|f'(\xi)| \leq |B_0'(\xi)|.$$

Proof. In view of (2.17) and (2.19) we have

$$\|B_0\|_{C(-1,1)} = \sqrt{\lambda}, \quad B_0(\pm b_j) = \pm(-1)^{j+1}\sqrt{\lambda}. \tag{3.38}$$

Thus the proposition is obvious for $\xi = \pm b_j$, $j = 1, 2, \dots$. For $f \in H_{\infty,R}$ and $\xi \neq \pm b_j$, $j = 1, 2, \dots$, put

$$Jf := \frac{1}{2\pi i} \int_{|z|=1} \frac{h(z) f(z)\, dz}{B_1(z)(z-\xi)^2(1-\xi z)^2},$$

where $h(z)$ and $B_1(z)$ are defined by (3.25). By Lemma 2.5, we get

$$Jf = \left(\frac{h(z)f(z)}{B_1(z)(1-\xi z)^2}\right)'\bigg|_{z=\xi} + \sum_{j=1}^{\infty} \frac{h(b_j)}{B_1'(b_j)} \left(\frac{f(b_j)}{(b_j-\xi)^2(1-\xi b_j)^2} - \frac{f(-b_j)}{(b_j+\xi)^2(1+\xi b_j)^2}\right).$$

Using (3.30), we have

$$f'(\xi) = -C(\xi)\left(\frac{h(z)}{B_1(z)(1-\xi z)^2}\right)'\bigg|_{z=\xi} f(\xi) + \frac{\pi C(\xi)}{2\sqrt{\lambda}\Lambda'(1-\lambda^2)} \times \sum_{j=1}^{\infty} (-1)^{j+1}(1-b_j^2)^2 \left(\frac{f(b_j)}{(b_j-\xi)^2(1-\xi b_j)^2} - \frac{f(-b_j)}{(b_j+\xi)^2(1+\xi b_j)^2}\right) + C(\xi)Jf, \tag{3.39}$$

where

$$C(\xi) := \frac{B_1(\xi)(1-\xi^2)^2}{h(\xi)}.$$

Putting

$$\psi_1(z) := \frac{z^2}{(z-\xi)^2(1-\xi z)^2},$$

and taking into account that $|B_0(e^{i\theta})| = 1$ almost everywhere, we have

$$Jf = \frac{1}{2\pi} \int_0^{2\pi} \overline{B_0(e^{i\theta})} \psi(e^{i\theta}) \psi_1(e^{i\theta}) f(e^{i\theta})\, d\theta,$$

where ψ is defined by (3.26). In Theorem 3.8 we proved that the function $\psi(e^{i\theta})$ is bounded and positive almost everywhere. Obviously, $\psi_1(e^{i\theta})$ has the same property. Thus, for all $f \in H_{\infty,R}$

$$|Jf| \leq \frac{1}{2\pi} \int_0^{2\pi} \psi(e^{i\theta})\psi_1(e^{i\theta})\, d\theta = JB_0. \tag{3.40}$$

Suppose that $C(\xi) > 0$. Then by (3.38), (3.39), and (3.40) we have

$$f'(\xi) \leq -C(\xi) \left(\frac{h(z)}{B_1(z)(1-\xi z)^2} \right)'\Bigg|_{z=\xi} f(\xi)$$
$$+ \frac{\pi C(\xi)}{2\Lambda'(1-\lambda^2)} \sum_{j=1}^{\infty} (1-b_j^2)^2 \left(\frac{1}{(b_j-\xi)^2(1-\xi b_j)^2} + \frac{1}{(b_j+\xi)^2(1+\xi b_j)^2} \right)$$
$$+ C(\xi) JB_0. \tag{3.41}$$

If $f(\xi) = B_0(\xi)$, then the right-hand side in the last inequality coincides with $B_0'(\xi)$ (see (3.39)). Thus

$$f'(\xi) \leq B_0'(\xi). \tag{3.42}$$

Put

$$g(z) := f\left(\frac{\alpha - z}{1 - \alpha z} \right), \qquad \alpha := \frac{2\xi}{1+\xi^2}.$$

Obviously, $g \in H_{\infty,R}$ and satisfies the condition (3.37). Moreover, $g(\xi) = f(\xi)$ and $g'(\xi) = -f'(\xi)$. Therefore, applying the inequality (3.42) to g, we obtain

$$-f'(\xi) \leq B_0'(\xi).$$

The case $C(\xi) < 0$ can be considered analogously by changing the inequality (3.41) to the opposite one. □

Denote by $H^{r,\beta}_{\infty,R}(R)$ the class of functions from $H^{r,\beta}_{\infty}(R)$ that are real on the real axis (for $r = 0$ the corresponding class is denoted by $H^{\beta}_{\infty,R}(R)$).

Proposition 3.13. *Let $f \in H^{\beta}_{\infty,R}(R)$ and let*

$$\|f\|_\infty \leq \|\phi_\lambda\|_\infty$$

for some $\lambda \in (0,1)$. If $a, b \in R$ and $f(a) = \phi_\lambda(b)$, then

$$|f'(a)| \leq |\phi_\lambda'(b)|.$$

Proof. Without loss of generality we can assume that $a = b$. Put

$$w(z) := \frac{4\beta}{\pi} \operatorname{arctanh} z.$$

This function maps conformally the unit disk D onto the strip S_β. Thus, $f(w(z)) \in H_{\infty,R}$. Since $\phi_\lambda(w(z)) = B_0(z)$ (see (2.16)), by putting $\xi = \tanh \frac{\pi a}{4\beta}$ in Proposition 3.12, we obtain

$$|f'(w(\xi))w'(\xi)| \leq |\phi'_\lambda(w(\xi))w'(\xi)|.$$

Hence, $|f'(a)| \leq |\phi'_\lambda(a)|$. □

From Proposition 3.13, applying the inductive scheme of the proof of the Kolmogorov comparison theorem, we obtain the following theorem.

Theorem 3.14. *Let $f \in H^{r,\beta}_{\infty,R}(R)$, $r \geq 0$, and let*

$$\|f\|_\infty \leq \|\Phi_{\lambda,r,\beta}\|_\infty \tag{3.43}$$

for some $\lambda \in (0,1)$. If $a, b \in R$ and $f(a) = \Phi_{\lambda,r,\beta}(b)$, then

$$|f'(a)| \leq |\Phi'_{\lambda,r,\beta}(b)|.$$

From Theorem 3.14 follows

Corollary 3.15. *Let $f \in H^{r,\beta}_{\infty,R}(R)$, $r \geq 0$, and let the inequality* (3.43) *holds for some $\lambda \in (0,1)$. Then for all $1 \leq k \leq r+1$*

$$\|f^{(k)}\|_\infty \leq \|\Phi^{(k)}_{\lambda,r,\beta}\|_\infty.$$

Remark 1. The statement of Corollary 3.15 remains valid for the class $H^{r,\beta}_\infty(R)$. Indeed, assume the converse. Then there exists $f_0 \in H^{r,\beta}_\infty(R)$ satisfying the inequality (3.43) for some $\lambda \in (0,1)$ and such that

$$\|f_0^{(k)}\|_\infty > \|\Phi^{(k)}_{\lambda,r,\beta}\|_\infty.$$

Consequently there exists $t_0 \in R$ for which

$$|f_0^{(k)}(t_0)| > \|\Phi^{(k)}_{\lambda,r,\beta}\|_\infty.$$

Without loss of generality we can assume that $f_0^{(k)}(t_0) > 0$. Consider the function

$$f_1(z) := \frac{f_0(z) + \overline{f_0(\overline{z})}}{2}.$$

It is obvious that $f_1 \in H^{r,\beta}_{\infty,R}(R)$, satisfies the inequality (3.43), and

$$f_1^{(k)}(t_0) = f_0^{(k)}(t_0) > \|\Phi^{(k)}_{\lambda,r,\beta}\|_\infty.$$

This contradicts Corollary 3.15.

Proof of Theorem 3.11. Put $\Phi_{\lambda,-1,\beta}(z) := \Phi'_{\lambda,0,\beta}(z)$. Then

$$\Phi^{(k)}_{\lambda,r,\beta}(z) = \Phi_{\lambda,r-k,\beta}(z), \quad k = 1, \dots, r+1.$$

It is easy to verify that the equalities (3.34) also hold for $r = -1$. Thus, using Remark 1, we see that it is sufficient to prove the existence of $\lambda \in (0,1)$ satisfying (3.36). Since

$$\|\Phi_{\lambda,0,\beta}\|_\infty = \sqrt{\lambda},$$

we have for $r = 0$, $\lambda = \delta^2$.

Let $r \geq 1$. Then from (3.34) it follows that $\|\Phi_{\lambda,r,\beta}\|_\infty$ depends continuously on λ. Since $\Lambda' \to \infty$ and $\Lambda \to \pi/2$ as $\lambda \to 0$, and (see (A.40))

$$\lim_{\lambda \to 0} \sqrt{\lambda} \exp\left(\frac{\pi\Lambda'}{4\Lambda}\right) = \frac{1}{2},$$

we get for all $j \geq 0$

$$\sqrt{\lambda} \sinh\left((2j+1)\frac{\pi\Lambda'}{2\Lambda}\right) \to \infty.$$

Thus $\|\Phi_{\lambda,r,\beta}\|_\infty \to 0$ as $\lambda \to 0$. If $\lambda \to 1$, then $\Lambda' \to \pi/2$ and $\Lambda \to \infty$. In this case it follows from (3.34) that $\|\Phi_{\lambda,r,\beta}\|_\infty \to \infty$. □

As it was yet mentioned, for $r = 0$ the equation (3.36) has the solution $\lambda = \delta^2$. Thus

$$\sup_{\substack{f \in H^\beta_\infty(R) \\ \|f\|_\infty \leq \delta}} \|f'\|_\infty = \frac{\delta\Lambda'}{2\beta}, \tag{3.44}$$

where Λ' is the complete elliptic integral of the first kind for the modulus $\lambda' = \sqrt{1-\delta^4}$ (the equality (3.44) is also follows from Corolarry 3.9). Denote by $\mathcal{H}^\beta_\infty(R)$ the space of functions which are analytic in the strip S_β and satisfy the condition

$$\|f\|_{\mathcal{H}^\beta_\infty(R)} := \sup_{z \in S_\beta} |f(z)| < \infty.$$

Then from (3.44) it follows that for all $f \in \mathcal{H}^{\beta}_{\infty}(R)$ the inequality

$$\|f'\|_\infty \le \frac{1}{2\beta}\|f\|_\infty \|f\|^2_{\mathcal{H}^{\beta}_{\infty}(R)} \int_0^{\pi/2} \frac{dx}{\sqrt{\|f\|^4_{\mathcal{H}^{\beta}_{\infty}(R)} \cos^2 x + \|f\|^4_\infty \sin^2 x}}$$

holds.

In contrast to the smooth case for the considered extremal problem on the class $H^{r,\beta}_{\infty}(R)$ we can pose the problem of norm estimation not only for intermediate derivatives but for derivatives of arbitrary orders. It turns out that for $k = r+2$ the equality (3.35) holds only for small δ. It was shown in Section 3.4 that for $\lambda = \delta^2$ the equation

$$\frac{1-5\lambda^2}{2}\left(\frac{\Lambda'}{\pi}\right)^2 = 1 \tag{3.45}$$

has the unique solution $\delta_0 = 0{,}2145\ldots$ (see (3.31). Put $\lambda_0 := \delta_0^2 = 0{,}0460\ldots$ and

$$\delta_r := \|\Phi_{\lambda_0,r,1}\|_\infty, \quad r \ge 1$$

(since $\|\Phi_{\lambda,r,\beta}\|_\infty = \sqrt{\lambda}$ the last equality is also valid for $r = 0$).

Theorem 3.16. *Let $r \ge 0$ and $\delta \in (0, \delta_r\beta^r]$. Then*

$$\sup_{\substack{f\in H^{r,\beta}_{\infty}(R)\\ \|f\|_\infty\le\delta}} \|f^{(r+2)}\|_\infty = \|\Phi''_{\lambda,0,\beta}\|_\infty = \left(\frac{\Lambda'}{2\beta}\right)^2 \sqrt{\lambda}(1-\lambda^2),$$

where λ satisfies (3.36).

Proof. In Theorem 3.10 we proved that for $0 < \lambda \le \lambda_0$

$$\sup_{\substack{f\in H^{\beta}_{\infty}(R)\\ \|f\|_\infty\le\sqrt{\lambda}}} \|f''\|_\infty = \|\Phi''_{\lambda,0,\beta}\|_\infty = \left(\frac{\Lambda'}{2\beta}\right)^2 \sqrt{\lambda}(1-\lambda^2). \tag{3.46}$$

It follows from (3.34) that

$$\|\Phi_{\lambda_0,r,\beta}\|_\infty = \delta_r\beta^r.$$

By analogy with the proof of Theorem 3.11, one can show that for $\delta \in (0, \delta_r\beta^r]$ there exists $\lambda \in (0, \lambda_0]$ satisfying (3.36). Now the assertion of the theorem follows from Theorem 3.11 and (3.46). □

Let us calculate the value δ_1. We have (see (A.46))

$$\|\Phi_{\lambda,1,1}\|_\infty = \int_0^{2\Lambda/\Lambda'} \sqrt{\lambda}\,\mathrm{sn}\left(\frac{\Lambda'}{2}z,\lambda\right)\,dz$$
$$= \frac{2\sqrt{\lambda}}{\Lambda'}\int_0^{\Lambda} \mathrm{sn}(t,\lambda)\,dt = \frac{1}{\sqrt{\lambda}\Lambda'}\log\frac{1+\lambda}{1-\lambda}.$$

Taking into account (3.45), we obtain

$$\delta_1 = \frac{1}{\pi}\left(\frac{1-5\lambda_0^2}{2\lambda_0}\right)^{1/2}\log\frac{1+\lambda_0}{1-\lambda_0} = 0{,}0961\ldots .$$

3.6 Notes and References

3.1. The problem of optimal recovery of the derivative of $f \in H_\infty$ was solved by Micchelli, Rivlin [1977]. The general case $1 \le p \le \infty$ was obtained by Osipenko, Stessin [1991]. An interesting minimization problem for optimal recovery of the derivative was considered by Rivlin, Ruscheweyh, Shaffer, Wirths [1983]. In this paper for a given $\xi \in (-1,0)$ an optimal recovery method of $f'(\xi)$, $f \in H_\infty$, from the knowledge of function values at points $z_1,\ldots,z_n \in [0,1)$, which are chosen to produce the least possible intrinsic error, is constructed. Previously Rivlin, Shaffer [1983] noted the unexpected fact that for $\xi = 0$ and $n = 1$ the intrinsic error is minimal if $z_1 = 1/\sqrt{3}$.

3.2. The material of this section is taken from Osipenko [1994a].

3.3. The material of this section is taken from Osipenko [1999].

3.4. Theorem 3.8 was proved by Osipenko, Stessin [1993]. The problem of optimal recovery of the second derivative which differs from the first derivative by a "switching" effect was studied in Osipenko [1994c].

3.5. There are many papers devoted to Kolmogorov-type inequalities for classes of smooth functions (see Tikhomirov, Magaril-Il'yaev [1985], Magaril-Il'yaev, Tikhomirov [1997]). The analogue of the classical Kolmogorov inequality for the Hardy–Sobolev class $H_\infty^{r,\beta}(R)$ considered in this section was obtained by Osipenko [1994b].

Chapter 4

Exact n-Widths of Analytic Functions

In Section 2.4 we looked for n optimal nodes to make the intrinsic error of optimal recovery minimal. Nevertheless, it is natural to ask: are there any n linear functionals which can be used instead of function values for which the error less? In other words, we would like to know what information is better to know about functions in order to recover them more precisely. This question leads to the notions of n-widths.

In Section 4.1 we find the exact values of the Kolmogorov, linear, Gel'fand, and information n-widths for the class H_p in $L_q(E,\mu)$, $1 \leq q \leq p \leq \infty$, where E is a subset of the unit disk. In Section 4.2 we obtain the exact order of $d_n(H_p, L_q(-1,1))$ for all $1 \leq q < p \leq \infty$. Section 4.3 is devoted to the periodic case. We calculate the exact values of n-widths for the Hardy–Sobolev classes in the case $p = \infty$.

In Section 4.4 we proved the Ismagylov type theorems for the linear, Gel'fand, and Bernstein n-widths in a Hilbert spaces. In Section 4.5 we apply these results to n widths of ellipsoids. Section 4.6 is devoted to the exact values of n-widths for the Hardy–Sobolev and Bergman–Sobolev classes for $p = 2$.

4.1 Exact n-Widths in H_p

Let X be a normed linear space. The *information n-width* of a set $W \in X$ is defined as follows

$$i_n(W,X) := \inf_{\substack{Y \supset W \\ l_1,\dots,l_n \in Y^*}} \inf_{S:\, M^n \to X} \sup_{f \in W} \|f - S(l_1 f,\dots,l_n f)\|_X, \tag{4.1}$$

where the infimum is taken over all normed linear spaces Y containing W; here $M = R$ or C. If the infimum in (4.1) is attained for some linear functionals, then these functionals are said to be *optimal* for corresponding information n-width.

If X is a normed linear space of functions defined on a set G and there exists $Y \supset W$ such that for all $t \in G$, $lf := f(t) \in Y^*$, then it is evident that

$$i_n(W,X) \le s_n(W,X). \tag{4.2}$$

For some classes of analytic functions we study whether the equality in (4.2) holds (that is, function values are optimal functionals) or not. If not, we shall be interested in optimal functionals which make possible to recover functions better than for function values. We shall also be interested in the connection between these quantities and the Kolmogorov, linear, and Gel'fand n-widths.

We start by recalling the definition of the various n-widths (see (1.3)). The Kolmogorov n-widths are defined by

$$d_n(W,X) := \inf_{X_n} \sup_{x \in W} \inf_{y \in X_n} \|x - y\|_X,$$

where X_n runs over all subspaces of X of dimension n or less.

The linear n-width is the quantity

$$\lambda_n(W,X) := \inf_{Y} \inf_{P_n} \sup_{x \in W} \|x - P_n x\|_X,$$

where Y are any normed linear spaces containing W and P_n are bounded linear operators mapping Y into X, whose range is n or less.

The Gel'fand n-width is defined as follows

$$d^n(W,X) := \inf_{Y} \inf_{Y^n} \sup_{x \in W \cap Y^n} \|x\|_X,$$

where Y is as in the definition of the linear n-width and Y^n are any subsets of W of codimension n (here we assume that $0 \in W$).

Much information on n-widths can be found in Pinkus [1985]. In particular, the following fundamental inequality holds:

$$d_n(W,X)\,,\; d^n(W,X) \leq \lambda_n(W,X). \tag{4.3}$$

Proposition 4.1. *Let W be a centrally symmetric set and $0 \in W$. Then*

$$d^n(W,X) \leq i_n(W,X) \leq \lambda_n(W,X). \tag{4.4}$$

Proof. The inequality

$$i_n(W,X) \leq \lambda_n(W,X)$$

evidently follows from the definition of information and linear n-widths. The lower bound we prove similarly to Proposition 2.9. Let $Y \supset W$ and $l_1, \ldots, l_n \in Y^*$. For each $\varepsilon > 0$ there exists $f_\varepsilon \in W$ such that $l_1 f_\varepsilon = \ldots = l_n f_\varepsilon = 0$ and

$$\sup_{\substack{f \in W \\ l_1 f = \ldots = l_n f = 0}} \|f\|_X \leq \|f_\varepsilon\|_X + \varepsilon.$$

For all algorithms S we have

$$\|f_\varepsilon - S(0,\ldots,0)\|_X + \|-f_\varepsilon - S(0,\ldots,0)\|_X \geq 2\|f_\varepsilon\|_X.$$

Therefore,

$$\sup_{f \in W} \|f - S(l_1 f, \ldots, l_n f)\|_X \geq \|f_\varepsilon\|_X \geq \sup_{\substack{f \in W \\ l_1 f = \ldots = l_n f = 0}} \|f\|_X - \varepsilon \geq d^n(W,X) - \varepsilon.$$

Taking the infimum over S and $l_1 \ldots, l_n$, we obtain

$$i_n(W,X) \geq d^n(W,X).$$

□

To calculate exact values of n-widths we shall frequently use the Borsuk Antipodality Theorem (Borsuk [1933]). For easy reference we state here a version of it.

Theorem 4.2. *Let Ω be a bounded, open, symmetric neighborhood of 0 in R^{n+1}, and T a continuous map of $\partial\Omega$ (the boundary of Ω) into R^n, with T odd on $\partial\Omega$, i.e., $T(-x) = -T(x)$ for all $x \in \partial\Omega$. Then there exists an $x^* \in \partial\Omega$ for which $T(x^*) = 0$.*

For the Hardy classes H_p there is a general result concerning exact values of n-widths which is formulated as follows.

Theorem 4.3. *Let E be a subset of the unit disk and μ be a positive measure on E such that the restriction operator, which maps $\mathcal{H}_p$ into $L_q := L_q(E,\mu)$, is compact. Then for all $1 \le q \le p \le \infty$*

$$d_n(H_p, L_q) = \lambda_n(H_p, L_q) = d^n(H_p, L_q)$$
$$= s_n(H_p, L_q) = i_n(H_p, L_q) = \inf_{B \in \mathcal{B}_n} \sup_{g \in H_p} \|Bg\|_{L_q},$$

where $\mathcal{B}_n$ is the set of all Blaschke products of degree at least n.

To prove this theorem we need some preliminary results relative to the extremal problem

$$\delta(p,q) := \sup_{g \in \mathcal{H}_p} \frac{\|g\|_{L_q}}{\|g\|_{\mathcal{H}_p}}. \tag{4.5}$$

A compactness argument shows the existence of solutions of (4.5). Moreover, we claim that any solution of this problem is free of zeros in the unit disk D. In fact, if $g \in \mathcal{H}_p$ has zero at a point $\alpha \in D$, then

$$g^*(z) := g(z) \Big/ \frac{z-\alpha}{1-\overline{\alpha}z} \in \mathcal{H}_p,$$

$\|g^*\|_{\mathcal{H}_p} = \|g\|_{\mathcal{H}_p}$ but $\|g^*\|_{L_q} > \|g\|_{L_q}$. We shall call a solution of (4.5) *normalized* if it has a $\mathcal{H}_p$ norm one and is positive at the point $z = 0$.

Proposition 4.4. *Let g be a normalized solution of* (4.5). *Then for all $1 \le q, p < \infty$ and all $\theta \in T$*

$$\delta^q(p,q)|g(e^{i\theta})|^p = \int_E |g(z)|^q P(e^{i\theta}, z)\, d\mu(z), \tag{4.6}$$

where

$$P(e^{i\theta}, z) = \frac{1-|z|^2}{|1-e^{-i\theta}z|^2}$$

is the Poisson kernel.

Proof. Let u be a real harmonic function on D which is continuous on $\overline{D}$. Let v denote the harmonic conjugate function of u. Set $g_\varepsilon := \exp(\varepsilon(u+iv))$, $\varepsilon \in R$. Since $\delta(p,q) \ge \|gg_\varepsilon\|_{L_q} / \|gg_\varepsilon\|_{\mathcal{H}_p}$, we have

$$\delta(p,q)\left(\frac{1}{2\pi}\int_0^{2\pi} |g(e^{i\theta})|^p e^{\varepsilon p u(e^{i\theta})}\, d\theta\right)^{1/p} \ge \left(\int_E |g(z)|^q e^{\varepsilon q u(z)}\, d\mu(z)\right)^{1/q}.$$

Expanding the exponentional terms and using the binomial theorem together with the fact that the solution g is normalized we obtain that

$$\begin{aligned}\delta^q(p,q)\frac{1}{2\pi}\int_0^{2\pi}|g(e^{i\theta})|^p u(e^{i\theta})\,d\theta &= \int_E |g(z)|^q u(z)\,d\mu(z)\\ &= \int_E |g(z)|^q \frac{1}{2\pi}\int_0^{2\pi} u(e^{i\theta})P(e^{i\theta},z)\,d\theta\,d\mu(z)\\ &= \frac{1}{2\pi}\int_0^{2\pi} u(e^{i\theta})\int_E |g(z)|^q P(e^{i\theta},z)\,d\mu(z)\,d\theta.\end{aligned}$$

Since $u(e^{i\theta})$ is an arbitrary continuous function on T, we have (4.6). □

Proposition 4.5. *For all* $1 \leq q \leq p \leq \infty$ *a normalized solution of* (4.5) *is uniquely determined.*

Proof. For $p = \infty$ it is evident that $g(z) \equiv 1$ is the unique normalized solution of (4.5). Assume that $p < \infty$. Let g_1 and g_2 be two normalized solutions of (4.5). It follows from (4.6) that

$$|g_1(e^{i\theta})/g_2(e^{i\theta})|^p = \int_E |g_1(z)/g_2(z)|^q\,d\nu(z),$$

where

$$d\nu(z) = \frac{|g_2(z)|^q P(e^{i\theta},z)\,d\mu(z)}{\int_E |g_2(z)|^q P(e^{i\theta},z)\,d\mu(z)}.$$

Since ν is a probability measure we have

$$|g_1(e^{i\theta})/g_2(e^{i\theta})|^p \leq \sup_{z\in E}|g_1(z)/g_2(z)|^q. \tag{4.7}$$

Put $u := \log|g_1/g_2|$. Both extremal functions g_1 and g_2 have no zeros in the unit disk. Consequently u is a harmonic function on D. It follows from (4.7) that

$$\sup_{\theta\in T} u(e^{i\theta}) \leq \frac{q}{p}\sup_{z\in E} u(z) \leq \sup_{z\in E} u(z).$$

The maximum principle implies that u is constant. Hence g_1 is a constant multiple of g_2 and the constant must be 1 since g_1 and g_2 are both normalized. □

Proof of Theorem 4.3. Let $z_0, \ldots, z_n$ be distinct points from the unit disk D, $w = (w_0, \ldots, w_n) \in C^{n+1}$, and $\sum_{j=0}^n |w_j|^2 = 1$. Put

$$\rho(w) := \inf_{\substack{f\in H_\infty\\ f(z_j)=w_j,\ j=0,\ldots,n}} \|f\|_{\mathcal{H}_\infty}.$$

According to the classical Pick–Nevanlinna Theorem (see, for example, Walsh [1960], Fisher [1983]) there is a unique Blashke product $B \in \mathcal{B}_n$ such that

$$\rho(w)B(z_j) = w_j, \quad j = 0, \dots, n. \tag{4.8}$$

This implies that the mapping

$$\alpha(w_0, \dots, w_n) = B,$$

where $w \in S^{2n+1}$ and B satisfies (4.8) is odd. Moreover, α is continuous in the topology of uniform convergence on compact subsets of D. Denote by g_B the unique normalized solution of (4.5) with respect to the measure $|B|^q \, d\mu$ when $q < \infty$. For $q = \infty$ put $g_B(z) \equiv 1$. Define β: $S^{2n+1} \to L_q$ by

$$\beta(w) := \alpha(w) g_{\alpha(w)}.$$

It is clear that β is a continuous odd mapping.

Let X_n be a complex n-dimensional subspace of L_q, $1 < q < \infty$, with basis $f_1, \dots, f_n$. For each $f \in L_q$ there exists the unique best approximation to f from X_n

$$f_0 = \sum_{j=1}^{n} c_j(f) f_j.$$

Put $\gamma(f) := (c_1(f), \dots, c_n(f))$. Then the mapping $\gamma \circ \beta$: $S^{2n+1} \to R^{2n}$ is continuous and odd. By the Borsuk Theorem (see Theorem 4.2) there exists a $w^* \in S^{2n+1}$ for which $c_j(\beta(w^*)) = 0$, $j = 1, \dots, n$. It means that there exists the Blashke product $B^* \in \mathcal{B}_n$ such that $c_j(B^* g_{B^*}) = 0$, $j = 1, \dots, n$. Thus

$$\begin{aligned} \sup_{f \in H_p} \inf_{h \in X_n} \|f - h\|_{L_q} &\geq \inf_{h \in X_n} \|B^* g_{B^*} - h\|_{L_q} = \|B^* g_{B^*}\|_{L_q} \\ &\geq \inf_{B \in \mathcal{B}_n} \|B g_B\|_{L_q} = \inf_{B \in \mathcal{B}_n} \sup_{g \in H_p} \|Bg\|_{L_q}. \end{aligned}$$

This completes the proof of the lower bound for the Kolmogorov n-width (the cases $q = 1, \infty$ are established by passing to the limit as either $q \searrow 1$ or $q \nearrow \infty$).

To obtain the lower bound for the Gel'fand n-width consider an arbitrary normed linear space Y containing $\mathcal{H}_p$ and arbitrary linear continuous functionals $l_1, \dots, l_n \in Y^*$. Denote by $\gamma_1 : S^{2n+1} \to C^n$ the mapping

$$\gamma_1(w) = (l_1(\beta(w)), \dots, l_n(\beta(w))).$$

γ_1 is an odd and continuous map. By the Borsuk Theorem there exists the Blashke product $B_0 \in \mathcal{B}_n$ such that $l_j(B_0 g_{B_0}) = 0$, $j = 1, \dots, n$. Thus

$$\sup_{\substack{f\in H_p \\ l_j(f)=0,\ j=1,\dots,n}} \|f\|_{L_q} \geq \|B_0 g_{B_0}\|_{L_q} \geq \inf_{B\in\mathcal{B}_n} \sup_{g\in H_p} \|Bg\|_{L_q}.$$

Now we prove the upper bound. In view of (4.2), (4.4), and ((4.3)) it suffices to prove it for $s_n(H_p, L_q)$ and $\lambda_n(H_p, L_q)$. For $p = \infty$ it follows from Theorem 2.1 that

$$s_n(H_\infty, L_q), \lambda_n(H_\infty, L_q) \leq \inf_{B\in\mathcal{B}_n} \|B\|_{L_q} = \inf_{B\in\mathcal{B}_n} \sup_{g\in H_\infty} \|Bg\|_{L_q}.$$

Let $p < \infty$. For $f \in H_p$, $\xi \in D$, and

$$B(z) = \prod_{j=1}^{k} \left(\frac{z - z_j}{1 - \bar{z}_j z}\right)^{\nu_j}$$

set

$$Jf(\xi) := B(\xi) g_B(\xi) \frac{1}{2\pi i} \int_{|z|=1} \frac{(1 - |\xi|^2) f(z)}{B(z) g_B(z)(z - \xi)(1 - \bar{\xi} z)}\, dz.$$

By the residue theorem we have

$$Jf(\xi) = f(\xi) - \sum_{j=1}^{k} \sum_{\nu=0}^{\nu_j - 1} c_{j\nu}(\xi) f^{(\nu_j)}(z_j)$$

($c_{j\nu}(\xi)$ may be determined exactly as it was done in Theorem 2.1). It is easy to check that

$$Jf(\xi) = B(\xi) g_B(\xi) \int_0^{2\pi} \frac{f(e^{i\theta})}{B(e^{i\theta}) g_B(e^{i\theta})}\, d\nu(\theta),$$

where

$$d\nu(\theta) = \frac{1}{2\pi} P(e^{i\theta}, \xi)\, d\theta.$$

Since ν is a probability measure, by the Hölder inequality we obtain

$$|Jf(\xi)|^q \leq |B(\xi) g_B(\xi)|^q \int_0^{2\pi} \left|\frac{f(e^{i\theta})}{g_B(e^{i\theta})}\right|^q d\nu(\theta).$$

Thus in view of (4.6)

$$
\begin{aligned}
\|Jf\|_{L_q}^q &\le \frac{1}{2\pi}\int_0^{2\pi}\left|\frac{f(e^{i\theta})}{g_B(e^{i\theta})}\right|^q \int_E |B(\xi)g_B(\xi)|^q P(e^{i\theta},\xi)\,d\mu(\xi)\,d\theta \\
&= \delta^q(p,q)\frac{1}{2\pi}\int_0^{2\pi}\left|\frac{f(e^{i\theta})}{g_B(e^{i\theta})}\right|^q |g_B(e^{i\theta})|^p\,d\theta \\
&= \delta^q(p,q)\frac{1}{2\pi}\int_0^{2\pi}|f(e^{i\theta})|^q|g_B(e^{i\theta})|^{p-q}\,d\theta.
\end{aligned}
$$

Applying the Hölder inequality to the latter integral with powers $p/q \ge 1$ and $(p/q)' = \dfrac{p}{p-q}$, we have

$$
\|Jf\|_{L_q}^q \le \delta^q(p,q)\left(\frac{1}{2\pi}\int_0^{2\pi}|f(e^{i\theta})|^p\,d\theta\right)^{\frac{q}{p}}\left(\frac{1}{2\pi}\int_0^{2\pi}|g_B(e^{i\theta})|^p\,d\theta\right)^{\frac{p-q}{p}} \le \delta^q(p,q).
$$

Consequently,

$$
\|Jf\|_{L_q} \le \delta(p,q) = \sup_{g\in H_p}\|Bg\|_{L_q}.
$$

As B is arbitrary we obtain the required estimates for $s_n(H_p, L_q)$ and $\lambda_n(H_p, L_q)$. □

Corollary 4.6. *For all* $1 \le q \le \infty$

$$
\begin{aligned}
d_n(H_\infty, L_q) &= \lambda_n(H_\infty, L_q) = d^n(H_\infty, L_q) \\
&= s_n(H_\infty, L_q) = i_n(H_\infty, L_q) = \inf_{B\in\mathcal{B}_n}\|B\|_{L_q}. \qquad (4.9)
\end{aligned}
$$

In Theorems 2.12, 2.16, and Corollary 2.17 we obtain a number of exact values of s_n which in view of Theorem 4.3 coincide with the Kolmogorov, linear, Gel'fand, and information n-widths. For example, we formulate the corresponding result for the n-widths of H_p in $L_q(T_\rho, \sigma_\rho)$.

Corollary 4.7. *For all* $1 \le q \le p \le \infty$ *and* $0 < \rho < 1$

$$
\begin{aligned}
d_n(H_p, L_q(T_\rho,\sigma_\rho)) &= \lambda_n(H_p, L_q(T_\rho,\sigma_\rho)) = d^n(H_p, L_q(T_\rho,\sigma_\rho)) \\
& s_n(H_p, L_q(T_\rho,\sigma_\rho)) = i_n(H_p, L_q(T_\rho,\sigma_\rho)) = \rho^n.
\end{aligned}
$$

4.2 Estimates of n-Widths for H_p in $L_q(-1,1)$

We apply the result of Theorem 4.3 in the case when $E = (-1,1)$ and μ is the Lebesgue measure. For $m \in N$ set

$$E_n(p,q,m) := \inf_{B \in \mathcal{B}_n} \sup_{g \in H_p} \|B^m g\|_{L_q(-1,1)}, \tag{4.10}$$

where $\mathcal{B}_n$ is the set of all Blaschke products of degree at least n. To obtain estimates for n-widths of the pair $(H_p, L_q(-1,1))$, we have to estimate $E_n(p,q,1)$.

First we prove some preliminary results.

Proposition 4.8. *For all $0 < \rho \leq 1$ and all $B \in \mathcal{B}_n$*

$$\int_{-\rho}^{\rho} \frac{\log|B(x)|}{1-x^2}\, dx \geq -n\frac{\pi^2}{4}.$$

Proof. It suffices to prove that for all z from the unit disk D

$$\int_{-\rho}^{\rho} \log\left|\frac{x-z}{1-\bar{z}x}\right| \frac{dx}{1-x^2} \geq -\frac{\pi^2}{4}.$$

We have

$$\begin{aligned}\int_{-\rho}^{\rho} \log\left|\frac{x-z}{1-\bar{z}x}\right| \frac{dx}{1-x^2} &\geq \int_{-1}^{1} \log\left|\frac{x-z}{1-\bar{z}x}\right| \frac{dx}{1-x^2} \\ &\geq \int_{-1}^{1} \log\left|\frac{x-\operatorname{Re} z}{1-\operatorname{Re} z\, x}\right| \frac{dx}{1-x^2} = \int_{-1}^{1} \log|u| \frac{du}{1-u^2} \\ &= 2\int_{0}^{1} \log|u| \frac{du}{1-u^2} = \int_{0}^{1} \log\frac{1-t}{1+t}\frac{dt}{t} = -\frac{\pi^2}{4}.\end{aligned}$$

The last integral can be calculated by using the Taylor series for the logarithmic function and the fact that

$$\sum_{k=1}^{\infty} \frac{1}{(2k-1)^2} = \frac{\pi^2}{8}.$$

□

Proposition 4.9. *For all $R > 0$ there exists a constant C_R such that for every $0 \leq r \leq R$ there are $a_1, \ldots, a_n > 0$ for which*

$$\max_{x \in [0,1]} x^r \prod_{k=1}^{n} \left|\frac{x-a_k}{x+a_k}\right| \leq C_R\, e^{-\pi\sqrt{nr}}. \tag{4.11}$$

Proof. We begin with the case $R \le 1$. For $r = 0$ the inequality (4.11) is evident. Assume that $0 < r \le 1$. Set

$$\varphi(u) := e^{\pi\sqrt{u/r}}, \quad y_k := \frac{\varphi(k)}{\varphi(m)}, \quad k = 1, \dots, m,$$

$$u(x) := \sum_{k=1}^{m} \log\left|\frac{x - y_k}{x + y_k}\right|.$$

Let us estimate the function $u(x)$ for $y_k < x < y_{k+1}$, $k = 1, \dots, m-1$. Put $\xi := x\varphi(m)$ and

$$g(u) := \log\left|\frac{\xi - \varphi(u)}{\xi + \varphi(u)}\right|.$$

Then

$$u(x) = v(\xi) := \sum_{k=1}^{m} g(k).$$

Since the function $g(u)$ is convex and negative for $u \in [1, k]$ and $u \in [k+1, m]$, we have

$$\int_1^k g(u)\,du \ge \sum_{j=1}^{k} g(j) - \frac{g(1) + g(k)}{2} \ge \sum_{j=1}^{k} g(j) - \frac{g(k)}{2},$$

$$\int_{k+1}^m g(u)\,du \ge \sum_{j=k+1}^{m} g(j) - \frac{g(k+1) + g(m)}{2} \ge \sum_{j=k+1}^{m} g(j) - \frac{g(k+1)}{2}.$$

Thus

$$\sum_{j=1}^{m} g(j) \le \int_1^m g(u)\,du - \int_k^{k+1} g(u)\,du + \frac{g(k) + g(k+1)}{2}.$$

We show that there is a constant C such that

$$\int_k^{k+1} g(u)\,du - \frac{g(k) + g(k+1)}{2} \ge C.$$

Making the substitution $y = \varphi(u)$ and then putting $y = \xi t$, we obtain

$$\int_k^{k+1} g(u)\,du = \frac{2r}{\pi^2}\int_{\varphi(k)}^{\varphi(k+1)} \log\left|\frac{\xi - y}{\xi + y}\right| \log y \,\frac{dy}{y}$$

$$= \frac{2r}{\pi^2}\int_{\varphi(k)/\xi}^{\varphi(k+1)/\xi} \log\left|\frac{1 - t}{1 + t}\right| \log \xi t \,\frac{dt}{t}.$$

Since $\varphi(k) < \xi < \varphi(k+1)$, there is a $\theta \in (0,1)$ such that $\xi = \exp(\pi\sqrt{(k+\theta)/r})$. Put

$$\begin{aligned} \delta &:= \frac{\varphi(k+1)}{\xi} - 1 = \pi\frac{1-\theta}{2\sqrt{kr}} + O\left(\frac{1}{k}\right), \\ \varepsilon &:= 1 - \frac{\varphi(k)}{\xi} = \pi\frac{\theta}{2\sqrt{kr}} + O\left(\frac{1}{k}\right). \end{aligned} \tag{4.12}$$

We have to prove that

$$\frac{2}{\pi}\sqrt{(k+\theta)r}\int_{1-\varepsilon}^{1+\delta} \log\left|\frac{1-t}{1+t}\right| \frac{dt}{t} + \frac{2r}{\pi^2}\int_{1-\varepsilon}^{1+\delta} \log\left|\frac{1-t}{1+t}\right| \frac{\log t}{t}\, dt - \frac{1}{2}\log\frac{\delta}{2+\delta} - \frac{1}{2}\log\frac{\varepsilon}{2-\varepsilon} \geq C.$$

Since

$$\frac{2}{\pi}\sqrt{(k+\theta)r} - \frac{1}{\delta+\varepsilon} = O(1),$$

it suffices to prove that

$$\frac{1}{\delta+\varepsilon}\int_{1-\varepsilon}^{1+\delta} \log\left|\frac{1-t}{1+t}\right| \frac{dt}{t} - \frac{1}{2}\log\frac{\delta}{2+\delta} - \frac{1}{2}\log\frac{\varepsilon}{2-\varepsilon} \geq C.$$

In view of the fact that the function

$$\frac{1-t}{t}\log\left|\frac{1-t}{1+t}\right|$$

is bounded it remains to show that

$$I := \frac{1}{\delta+\varepsilon}\int_{1-\varepsilon}^{1+\delta} \log\left|\frac{1-t}{1+t}\right| dt - \frac{1}{2}\log\frac{\delta}{2+\delta} - \frac{1}{2}\log\frac{\varepsilon}{2-\varepsilon} \geq C.$$

Taking into account (4.12), we find

$$I = \frac{1-2\theta}{\theta}\log\frac{1-\theta}{\theta} + O\left(\frac{1}{\sqrt{k}}\right).$$

Thus we proved that there is a constant C such that for all $x \in [y_1, y_m]$

$$\sum_{j=1}^{m} g(j) \leq \int_1^m g(u)\, du - C. \tag{4.13}$$

Denote the latter integral by J. We have

$$J = \frac{2r}{\pi^2} \int_{\xi^{-1}\varphi(1)}^{x^{-1}} \log \left| \frac{1-t}{1+t} \right| \log \xi t \, \frac{dt}{t} \le C + \frac{2r}{\pi^2} \log \xi \int_0^\infty \log \left| \frac{1-t}{1+t} \right| \frac{dt}{t}$$
$$+ \frac{2r}{\pi^2} \log \xi \left(\int_{x^{-1}}^\infty \log \left| \frac{1+t}{1-t} \right| \frac{dt}{t} + \int_0^{\xi^{-1}\varphi(1)} \log \left| \frac{1+t}{1-t} \right| \frac{dt}{t} \right).$$

Using that

$$\int_0^\infty \log \left| \frac{1-t}{1+t} \right| \frac{dt}{t} = -\frac{\pi^2}{2}$$

and for all $x \in [0,1]$

$$\int_0^x \log \left| \frac{1-t}{1+t} \right| \frac{dt}{t} \le 4x,$$

we get

$$J \le C - r \log \xi + \frac{2r}{\pi^2} \log \xi \left(4x + 4\xi^{-1}\varphi(1) \right) \le C - r \log \xi + \frac{8}{\pi} x \sqrt{mr}.$$

Since

$$x \le \frac{1}{2} \log \left| \frac{x+1}{x-1} \right|,$$

we have

$$J \le C - r \log x - \pi\sqrt{mr} + 2[\sqrt{m}] \log \left| \frac{x+1}{x-1} \right|.$$

From here and (4.13)

$$\sum_{k=1}^m \log \left| \frac{x - y_k}{x + y_k} \right| + 2[\sqrt{m}] \log \left| \frac{x-1}{x+1} \right| \le C - r \log x - \pi\sqrt{mr}.$$

Putting $n_m := m + 2[\sqrt{m}]$ and observing that $\sqrt{mr} \ge \sqrt{n_m r} - 1$, we find that the rational function

$$f(x) := \prod_{k=1}^m \frac{x - y_k}{x + y_k} \left(\frac{x-1}{x+1} \right)^{2[\sqrt{m}]}$$

for all $[y_1, 1]$ satisfies the inequality

$$x^r |f(x)| \le C e^{-\pi\sqrt{n_m r}}.$$

For $x \in [0, y_1]$, taking into account that $|f(x)| \le 1$, we have

$$x^r |f(x)| \le y_1^r \le C e^{-\pi\sqrt{n_m r}}.$$

Since $n_{m+1}-n_m \le 3$ we proved the assertion of the proposition for all $n \in N$ and $R \le 1$.

Assume now that $R > 1$ and $s \le R$. Put $r = s/([R]+1)$. Since $r < 1$ there exists a rational function

$$f(x) = \prod_{j=1}^{m} \frac{x-a_j}{x+a_j}$$

such that

$$x^r |f(x)| \le C e^{-\pi\sqrt{mr}}.$$

Consequently,

$$x^{r([R]+1)} |f(x)|^{[R]+1} \le C^{[R]+1} e^{-\pi\sqrt{mr}([R]+1)}.$$

Thus we have found a rational function $f_1(x) = f^{[R]+1}(x)$ of degree $n_m = m([R]+1)$ such that

$$x^s |f_1(x)| \le C^{[R]+1} e^{-\pi\sqrt{n_m s}}.$$

Since $n_{m+1} - n_m \le [R]+1$ the assertion of the proposition is proved for all $n \in N$. □

By the mappings

$$t = \pm\frac{x-1}{x+1}$$

we obtain

Corollary 4.10. *For all $R > 0$ there exists a constant C_R such that for every $0 \le r \le R$ there is a $B \in \mathcal{B}_n$ for which*

$$\max_{x\in[-1,1]} (1-x^2)^r |B(x)| \le C_R e^{-\pi\sqrt{nr/2}}.$$

Now we can prove the main result of this section.

Theorem 4.11. *For all $1 \le q \le p \le \infty$ and $m \in N$ there exist constants C_1 and C_2 such that*

$$C_1 n^{(1/q-1/p)/2} \exp\left(-\pi\sqrt{\frac{nm}{2}\left(\frac{1}{q}-\frac{1}{p}\right)}\right) \le E_n(p,q,m)$$
$$\le C_2 n^{(1/q-1/p)/2} \exp\left(-\pi\sqrt{\frac{nm}{2}\left(\frac{1}{q}-\frac{1}{p}\right)}\right).$$

Proof. We start with the lower estimate. Set

$$\varphi(z) := \frac{1}{(1-z^2)^{1/p-\varepsilon}}$$

where $0 < \varepsilon \leq 1$. We have

$$\|\varphi\|_{\mathcal{H}_p}^p = \frac{2^{\varepsilon p}}{\pi} \int_0^{\pi/2} \sin^{\varepsilon p - 1} \theta \, d\theta \leq \frac{\pi^{\varepsilon p}}{2\varepsilon p}.$$

Consequently, for $C_0 := \pi^{-1}(2p)^{1/p}$

$$\psi(z) := C_0 \, \varepsilon^{1/p} \varphi(z) \in H_p.$$

For $0 < \rho < 1$ set

$$C_\rho := \log^{-1} \frac{1+\rho}{1-\rho}, \quad w(x) := \frac{C_\rho}{1-x^2}.$$

Since

$$\int_{-\rho}^{\rho} w(x) \, dx = 1,$$

Jensen's inequality implies that for every $B \in \mathcal{B}_n$

$$\begin{aligned} I_B := \|B\varphi\|_{L_q(-1,1)}^q &\geq \int_{-\rho}^{\rho} \frac{|B(x)|^q}{(1-x^2)^{q/p-\varepsilon q}} w^{-1}(x) w(x) \, dx \\ &\geq \int_{-\rho}^{\rho} |B(x)|^q w^{-1+q/p-\varepsilon q}(x) C_\rho^{-q/p+\varepsilon q} w(x) \, dx \\ &\geq \exp \int_{-\rho}^{\rho} \left(q \log |B(x)| - \left(1 - \frac{q}{p} + \varepsilon q\right) \log w(x) \right. \\ &\qquad \left. - \left(\frac{q}{p} - \varepsilon q\right) \log C_\rho \right) w(x) \, dx. \end{aligned}$$

It follows from Proposition 4.8 that

$$\int_{-\rho}^{\rho} q \log |B(x)| w(x) \, dx \geq -C_\rho q n \frac{\pi^2}{4}.$$

Furthermore, we have

$$\begin{aligned} \int_{-\rho}^{\rho} \log w(x) w(x) \, dx &= \log C_\rho - 2C_\rho \int_0^{\rho} \frac{\log(1-x^2)}{1-x^2} \, dx \\ &\leq \log C_\rho - C_\rho \int_0^{\rho} \log(1-x) \left(\frac{1}{1+x} + \frac{1}{1-x} \right) dx \\ &\leq \log C_\rho + C_\rho \left(1 + \frac{1}{2} \log^2(1-\rho) \right). \end{aligned}$$

Observing that

$$C_\rho \leq \frac{1}{-\log(1-\rho)},$$

we obtain

$$\begin{aligned} I_B &\geq \exp\left[-C_\rho q n\frac{\pi^2}{4} - \log C_\rho - C_\rho\left(1-\frac{q}{p}+\varepsilon q\right)\left(1+\frac{1}{2}\log^2(1-\rho)\right)\right] \\ &\geq \exp\left[\frac{1}{\log(1-\rho)}qn\frac{\pi^2}{4} + \log\left(-\log(1-\rho)\right)\right. \\ &\qquad \left. + \left(1-\frac{q}{p}+\varepsilon q\right)\left(\frac{1}{\log(1-\rho)}+\frac{1}{2}\log(1-\rho)\right)\right]. \end{aligned}$$

Choosing the parameter ρ such that

$$\log(1-\rho) = -\sqrt{n}\frac{\pi}{\sqrt{2}}\frac{1}{\sqrt{1/q-1/p+\varepsilon}},$$

we have

$$I_B \geq \exp\left(-\pi q\sqrt{\frac{n}{2}\left(\frac{1}{q}-\frac{1}{p}+\varepsilon\right)} + \log\sqrt{n} + C\right).$$

Thus

$$\begin{aligned} E_n(p,q,1) &\geq \inf_{B\in\mathcal{B}_n}\|B\psi\|_{L_q(-1,1)} = C_0\varepsilon^{1/p}\inf_{B\in\mathcal{B}_n} I_B^{1/q} \\ &\geq C\varepsilon^{1/p}n^{1/(2q)}\exp\left(-\pi\sqrt{\frac{n}{2}\left(\frac{1}{q}-\frac{1}{p}+\varepsilon\right)}\right) \\ &\geq C\varepsilon^{1/p}n^{1/(2q)}\exp\left(-\pi\sqrt{\frac{n}{2}\left(\frac{1}{q}-\frac{1}{p}\right)} + C\sqrt{n}\varepsilon\right). \end{aligned}$$

Putting $\varepsilon = 1/\sqrt{n}$, we get the lower estimate for $m=1$. The case $m>1$ follows from the fact that

$$E_n(p,q,m) \geq E_{nm}(p,q,1).$$

Now we obtain the upper estimate. If $p=q$, then

$$E_n(p,p,m) \leq \sup_{g\in H_p}\|g\|_{L_p(-1,1)}.$$

The Fejér–Riesz inequality (see Duren [1970, . 46]) yields

$$\|g\|_{L_p(-1,1)} \leq \pi^{1/p}\|g\|_{\mathcal{H}_p} \leq \pi^{1/p}.$$

Thus

$$E_n(p,p,m) \le \pi^{1/p}.$$

For $p > q$ we set $s := p/q > 1$ and apply the Hölder inequality

$$E_n(p,q,m) \le \sup_{g\in H_p} \inf_{B\in\mathcal{B}_n} \left(\int_{-1}^{1} |g(x)|^{qs}\,dx\right)^{1/(qs)} \left(\int_{-1}^{1} |B(x)|^{mqs'}\,dx\right)^{1/(qs')}.$$

Putting

$$t := mqs' = m\left(\frac{1}{q} - \frac{1}{p}\right)^{-1}, \tag{4.14}$$

we obtain

$$E_n(p,q,m) \le \sup_{g\in H_p} \inf_{B\in\mathcal{B}_n} \|g\|_{L_p(-1,1)} \|B\|^m_{L_t(-1,1)} \le \pi^{1/p} \inf_{B\in\mathcal{B}_n} \|B\|^m_{L_t(-1,1)}.$$

Let $r := (1 - n^{-1/2})/t$. From Proposition 4.9 we have

$$\begin{aligned}\|B\|_{L_t(-1,1)} &\le \left(\int_{-1}^{1} (1-x^2)^{-rt}\,dx\right)^{1/t} \max_{t\in[-1,1]} (1-x^2)^r |B(x)| \\ &\le Cn^{1/(2t)} \exp\left(-\pi\sqrt{\frac{n}{2t}}\right).\end{aligned}$$

Taking into account (4.14), we get the upper estimate. □

From Theorems 4.3 and 4.11 immediately follows

Theorem 4.12. *For all $1 \le q < p \le \infty$ and the pair $(H_p, L_q(-1,1))$ there exist constants C_1 and C_2 such that*

$$\begin{aligned}C_1 n^{(1/q-1/p)/2} \exp\left(-\pi\sqrt{\frac{n}{2}\left(\frac{1}{q}-\frac{1}{p}\right)}\right) &\le d_n = \lambda_n = d^n = s_n = i_n \\ &\le C_2 n^{(1/q-1/p)/2} \exp\left(-\pi\sqrt{\frac{n}{2}\left(\frac{1}{q}-\frac{1}{p}\right)}\right).\end{aligned}$$

4.3 Exact n-Widths of Periodic Functions

Denote by $H^{r,\beta}_{\infty,R}(T)$ the set of all 2π-periodic real-valued functions that can be analytically continued in the strip S_β so that $|f^{(r)}(z)| \le 1$ (for $r = 0$ the corresponding class we denote by $H^{\beta}_{\infty,R}(T)$). To find exact values of n-widths on the class $H^{r,\beta}_{\infty,R}(T)$ we need several preliminary propositions.

Let φ be a real-valued continuous function defined on $[-1,1]$. For a real-valued continuous 2π-periodic function $\mathcal{K}$ we set

$$\Theta_{2n}(\varphi,\mathcal{K}) := \{\,\eta = (\eta_1,\dots,\eta_{2n}) : 0 \le \eta_1 \le \dots \le \eta_{2n} < 2\pi,\ \varphi(\mathcal{K} * h_\eta) \perp 1\,\},$$

where

$$h_\eta(t) := (-1)^j, \quad t \in [\eta_{j-1},\eta_j), \quad j = 1,\dots,2n+1.$$

Denote by D_r the Bernouilli Monospline:

$$D_r(t) := \frac{1}{\pi}\sum_{k=1}^{\infty}\frac{\cos(kt - \pi r/2)}{k^r}, \quad r = 1,2,\dots\ .$$

Lemma 4.13. *Let φ be a continuous, odd, and strictly increasing function defined on $[-1,1]$. Assume that $\mathcal{K} \in NCVD$. Then*

$$\inf_{\substack{a\in R\\ \eta\in\Theta_{2n}(\varphi,\mathcal{K})}} \|a + D_r * \varphi(\mathcal{K} * h_\eta)\|_\infty = \|D_r * \varphi(\mathcal{K} * h_n)\|_\infty.$$

Proof. Let $\eta \in \Theta_{2n}(\varphi,\mathcal{K})$. Set

$$f_\eta := D_r * \varphi(\mathcal{K} * h_\eta), \qquad f_n := D_r * \varphi(\mathcal{K} * h_n).$$

Suppose that there exists an $a \in R$ and a $\eta \in \Theta_{2n}(\varphi,\mathcal{K})$ for which

$$\|a + f_\eta\|_\infty < \|f_n\|_\infty.$$

As $f_n(t+\pi/n) = -f_n(t)$, there exists at least $2n$ points $0 \le t_1 < \dots < t_{2n} < 2\pi$ such that

$$f_n(t_j) = \varepsilon(-1)^j\|f_n\|_\infty, \quad j = 1,\dots,2n,$$

where $\varepsilon = 1$ or -1. Denote by $Z(f)$ the number of distinct zeros of a function f on T. Then for any $\alpha \in T$ we have

$$Z\Big(f_n(\cdot + \alpha) - a - f_\eta(\cdot)\Big) \ge 2n.$$

By Rolle's theorem

$$S\left(f_n^{(r)}(\cdot+\alpha)-f_\eta^{(r)}(\cdot)\right)=S\left(\varphi((\mathcal{K}*h_n)(\cdot+\alpha))-\varphi((\mathcal{K}*h_\eta)(\cdot))\right)\geq 2n.$$

Since φ is an odd and strictly increasing function,

$$S\left((\mathcal{K}*h_n)(\cdot+\alpha)-(\mathcal{K}*h_\eta)(\cdot)\right)\geq 2n.$$

Taking into account that $\mathcal{K}\in NCVD$, we have

$$S(h_n(\cdot+\alpha)-h_\eta(\cdot))\geq 2n.$$

Choosing $\alpha=-\eta_1$, it is easy to verify that

$$S(h_n(\cdot-\eta_1)-h_\eta(\cdot))\leq 2(n-1).$$

This contradiction proves the lemma. □

Set $\varphi_0(z):=\tan\dfrac{\pi}{4}z$,

$$\Phi_{n,0}^\beta:=\varphi_0(\mathcal{K}_\beta*h_n),\qquad \Phi_{n,r}^\beta:=D_r*\varphi_0(\mathcal{K}_\beta*h_n),\quad r=1,2,\dots\ ,$$

where

$$\mathcal{K}_\beta(z)=\frac{1}{2\pi}\left(1+2\sum_{j=1}^{\infty}\frac{\cos jz}{\cosh j\beta}\right).$$

It follows from (2.28) that

$$\Phi_{n,0}^\beta(z)=\sqrt{\lambda}\,\mathrm{sn}\left(\frac{2n\Lambda}{\pi}z,\lambda\right),$$

where λ such that

$$\frac{\pi\Lambda'}{2\Lambda}=2n\beta$$

(in other words, $\lambda=\kappa(2\beta n)$ where κ is defined by (2.5)).

Thus for $\lambda=\kappa(2\beta n)$ the functions $\Phi_{n,r}^\beta$ coincide with the functions $\Phi_{\lambda,r,\beta}$ introduced in Section 3.5. In view of (3.34) we have

$$\begin{aligned}\Phi_{n,r}^\beta(t)&=\frac{\pi}{\sqrt{\lambda}\Lambda n^r}\sum_{s=0}^{\infty}\frac{\sin((2s+1)nt-\pi r/2)}{(2s+1)^r\sinh((2s+1)2n\beta)},\\ \|\Phi_{n,r}^\beta\|_\infty&=\frac{\pi}{\sqrt{\lambda}\Lambda n^r}\sum_{s=0}^{\infty}\frac{(-1)^{s(r+1)}}{(2s+1)^r\sinh((2s+1)2n\beta)},\end{aligned}\qquad r=0,1,\dots\ .$$

Lemma 4.14. *For all* $r \geq 0$

$$\sup_{\substack{f \in H^{r,\beta}_{\infty,R}(T) \\ c_j(f)=0,\ |j|<n}} |f(0)| = \|\Phi^{\beta}_{n,r}\|_{\infty},$$

where

$$c_j(f) = \frac{1}{2\pi}\int_0^{2\pi} f(t)e^{-ijt}\,dt.$$

Proof. Put

$$\psi(t) := \begin{cases} \Phi^{\beta}_{n,r}(t), & r = 2k-1, \\ \Phi^{\beta}_{n,r}\left(t + \dfrac{\pi}{2n}\right), & r = 2k. \end{cases}$$

Note that ψ is an even function and $|\psi(0)| = \|\Phi^{\beta}_{n,r}\|_{\infty}$. Suppose there exists a function $f_0 \in H^{r,\beta}_{\infty,R}(T)$ such that $c_j(f_0) = 0$, $|j| < n$, and $|f_0(0)| > \|\Phi^{\beta}_{n,r}\|_{\infty}$. Set

$$\rho := \frac{\psi(0)}{f_0(0)}, \qquad g_0(t) := \frac{f_0(t) + f_0(-t)}{2}, \qquad F := \psi - \rho g_0.$$

Since the trigonometric polynomials of degree at most $n-1$ are a Chebyshev system of dimension $2n-1$, it follows from the equalities $c_j(F) = 0$, $|j| < n$, that F has at least $2n$ cyclic sign changes (see Pinkus [1985, p. 41]). In addition F has a zero at $z = 0$. As F is an even function the number of distinct zeros of F at least $2n+1$. By Rolle's theorem the same conclusion remains valid for $F^{(r)}$.

By means the mapping $w = \dfrac{1}{i}\log z$, we see that $F^{(r)}\left(\dfrac{1}{i}\log z\right)$ has at least $2n+1$ zeros in the annulus $\Delta_\beta := \{z \in C : e^{-\beta} < |z| < e^{\beta}\}$. It follows from the equality

$$\psi^{(r)}\left(\frac{1}{i}\log z\right) = \begin{cases} \sqrt{\lambda}\,\mathrm{sn}\left(\dfrac{2n\Lambda}{\pi i}\log z, \lambda\right), & r = 2k-1, \\ \sqrt{\lambda}\,\mathrm{sn}\left(\dfrac{2n\Lambda}{\pi i}\log z + \Lambda, \lambda\right), & r = 2k, \end{cases}$$

that $\psi^{(r)}\left(\dfrac{1}{i}\log z\right)$ has exactly $2n$ zeros in the annulus Δ_β. Since for $z \in \partial\Delta_\beta$

$$\left|\psi^{(r)}\left(\frac{1}{i}\log z\right) - F^{(r)}\left(\frac{1}{i}\log z\right)\right| = \left|\rho g_0^{(r)}\left(\frac{1}{i}\log z\right)\right|$$
$$\leq |\rho| < 1 \equiv \left|\psi^{(r)}\left(\frac{1}{i}\log z\right)\right|,$$

Rouche's theorem implies that $F^{(r)}\left(\frac{1}{i}\log z\right)$ has at least $2n$ zeros. This contradiction proves that

$$|f(0)| \le \|\Phi^{\beta}_{n,r}\|_\infty \tag{4.15}$$

for any function $f \in H^{r,\beta}_{\infty,R}(T)$ such that $c_j(f) = 0$, $|j| < n$. As $\psi \in H^{r,\beta}_{\infty,R}(T)$ the inequality (4.15) is sharp. □

Theorem 4.15. *For all $r \ge 0$ and the pair $(H^{r,\beta}_{\infty,R}(T), L_\infty(T))$*

$$d_{2n} = \lambda_{2n} = d^{2n} = i_{2n} = d_{2n-1} = \lambda_{2n-1} = d^{2n-1} = i_{2n-1} = \|\Phi^{\beta}_{n,r}\|_\infty .$$

Moreover, for $r = 0$ and $1 \le q \le \infty$ for the pair $(H^{\beta}_{\infty,R}(T), L_q(T))$

$$d_{2n} = \lambda_{2n} = d^{2n} = i_{2n} = s_{2n} = \|\Phi^{\beta}_{n,0}\|_q$$
$$= \begin{cases} \sqrt{\lambda}\left(\dfrac{2\pi}{\Lambda} J_q(\lambda)\right)^{1/q}, & 1 \le q < \infty, \\ \sqrt{\lambda}, & q = \infty, \end{cases}$$

where $\lambda = \kappa(2\beta n)$ and

$$J_q(\lambda) := \int_0^1 \frac{t^q\,dt}{\sqrt{(1-t^2)(1-\lambda^2 t^2)}}.$$

For $i_{2n-1}(H^{r,\beta}_{\infty,R}(T), L_\infty(T))$ and $i_{2n}(H^{r,\beta}_{\infty,R}(T), L_\infty(T))$ the Fourier coefficients

$$a_j(f) := \frac{1}{2\pi}\int_0^{2\pi} f(t)\cos jt\,dt, \quad j = 0, \dots, n-1,$$
$$b_j(f) := \frac{1}{2\pi}\int_0^{2\pi} f(t)\sin jt\,dt, \quad j = 1, \dots, n-1.$$

are optimal functionals. Samling at $2n$ equidistant points on T is optimal for $i_{2n}(H^{\beta}_{\infty,R}(T), L_q(T))$.

Proof. We will first prove the lower bound for the Kolmogorov and Gel'fand $2n$-widths. Set

$$S^{2n} := \left\{ x = (x_1, \dots, x_{2n+1}) \in R^{2n+1} \,\middle|\, \sum_{s=1}^{2n+1} |x_s| = 2\pi \right\},$$

$$\tau_0(x) := 0, \qquad \tau_j(x) := \sum_{s=1}^{j} |x_s|, \quad j = 1, \dots, 2n+1.$$

For each $x \in S^{2n}$ put

$$g_x(t) := \operatorname{sign} x_j, \quad \tau_{j-1}(x) \le t < \tau_j(x), \quad j = 1, \dots, 2n+1,$$

$$f_x := \begin{cases} \varphi_0(\mathcal{K}_\beta * g_x), & r = 0, \\ D_r * \varphi_0(\mathcal{K}_\beta * g_x), & r \ge 1. \end{cases}$$

Let X_{2n} be any $2n$-dimensional subspace of L_q, $1 < q < \infty$. Suppose that $X_{2n} = \operatorname{span}\{f_1, \dots, f_{2n}\}$ and

$$f_x^* := \sum_{j=1}^{2n} \alpha_j(x) f_j$$

is the best approximation to f_x from X_{2n}. Consider the mapping

$$\alpha(x) := \begin{cases} (\alpha_1(x), \alpha_2(x), \dots, \alpha_{2n}(x)), & r = 0, \\ (b(x), \alpha_2(x), \dots, \alpha_{2n}(x)), & r \ge 1, \end{cases}$$

where

$$b(x) := \int_0^{2\pi} \varphi_0\Big((\mathcal{K}_\beta * g_x)(t)\Big)\, dt.$$

The mapping $\alpha\colon S^{2n} \to R^{2n}$ is continuous and odd. Therefore by Borsuk's theorem there exists an $x^* \in S^{2n}$ for which $\alpha(x^*) = 0$. As the function $\varphi_0(z)$ maps the strip $|\operatorname{Re} z| < 1$ conformally onto the unit disk D, $f_{x^*} \in H^{r,\beta}_{\infty,R}(T)$.

If $r = 0$, then by Theorem 2.13 we have

$$\sup_{f \in H^{\beta}_{\infty,R}(T)} \inf_{g \in X_{2n}} \|f - g\|_q \ge \inf_{g \in X_{2n}} \|f_{x^*} - g\|_q = \|f_{x^*}\|_q$$

$$\ge \inf_{\eta \in \Theta_{2n}} \|\varphi_0(\mathcal{K}_\beta * h_\eta)\|_q = \|\varphi_0(\mathcal{K}_\beta * h_n)\|_q = \|\Phi^{\beta}_{n,0}\|_q.$$

Consequently,

$$d_{2n}(H^{\beta}_{\infty,R}(T), L_q(T)) \ge \|\Phi^{\beta}_{n,0}\|_q.$$

The cases $q = 1, \infty$ are established by passing to the limit as either $q \searrow 1$ or $q \nearrow \infty$.

Let $r \ge 1$ and $1 \in X_{2n}$ (assume that $f_1(t) \equiv 1$). Then

$$\sup_{f \in H^{r,\beta}_{\infty,R}(T)} \inf_{g \in X_{2n}} \|f - g\|_q \ge \inf_{g \in X_{2n}} \|f_{x^*} - g\|_q = \|f_{x^*} - \alpha_1(x^*)\|_q$$

$$\ge \inf_{\substack{a \in R \\ \eta \in \Theta_{2n}(\varphi_0, \mathcal{K}_\beta)}} \|a + D_r * \varphi_0(\mathcal{K}_\beta * h_\eta)\|_q. \tag{4.16}$$

If $1 \notin X_{2n}$, then the left-hand side of (4.16) is equal to $+\infty$. Thus,

$$d_{2n}(H^{r,\beta}_{\infty,R}(T), L_q(T)) \geq \inf_{\substack{a\in R \\ \eta\in\Theta_{2n}(\varphi_0,\mathcal{K}_\beta)}} \|a + D_r * \varphi_0(\mathcal{K}_\beta * h_\eta)\|_q.$$

By passing to the limit $q \to \infty$ using Lemma 4.13, we obtain

$$d_{2n}(H^{r,\beta}_{\infty,R}(T), L_\infty(T)) \geq \|D_r * \varphi_0(\mathcal{K}_\beta * h_n)\|_\infty = \|\Phi^{\beta}_{n,r}\|_\infty.$$

Let us now prove the lower bound for the Gel'fand $2n$-widths. Let Y be any normed linear space containing $H^{r,\beta}_{\infty,R}(T)$ and

$$X^{2n} = \{ f \in Y : \langle l_j, f\rangle = 0,\ j = 1, \dots, 2n,\ l_j \in Y^* \}.$$

Consider the case $r = 0$. The mapping

$$\alpha(x) := (\langle l_1, f_x\rangle, \dots, \langle l_{2n}, f_x\rangle)$$

is continuous and odd. By Borsuk's theorem there exists an $x^* \in S^{2n}$ for which $\alpha(x^*) = 0$. Consequently,

$$\sup_{f\in H^{\beta}_{\infty,R}(T)\cap X^{2n}} \|f\|_q \geq \|f_{x^*}\|_q \geq \inf_{\eta\in\Theta_{2n}} \|\varphi_0(\mathcal{K}_\beta * h_\eta)\|_q = \|\Phi^{\beta}_{n,0}\|_q.$$

Hence,

$$d^{2n}(H^{\beta}_{\infty,R}(T), L_q(T)) \geq \|\Phi^{\beta}_{n,0}\|_q$$

for all $1 \leq q \leq \infty$.

Let $r \geq 1$. If $\langle l_j, 1\rangle = 0$, $j = 1, \dots, 2n$, then

$$\sup_{f\in H^{r,\beta}_{\infty,R}(T)\cap X^{2n}} \|f\|_\infty = \infty.$$

Assume that $\langle l_1, 1\rangle \neq 0$. Set

$$L_j := l_j - \frac{\langle l_j, 1\rangle}{\langle l_1, 1\rangle} l_1, \quad j = 2, \dots, 2n.$$

Denote by α the mapping

$$\alpha(x) := (b(x), \langle L_2, f_x\rangle, \dots, \langle L_{2n}, f_x\rangle).$$

Since α: $S^{2n} \to R^{2n}$ is an odd and continuous map, by Borsuk's theorem there exists an $x^* \in S^{2n}$ for which $\alpha(x^*) = 0$. Then

$$f^* := f_{x^*} - \frac{\langle l_1, f_{x^*}\rangle}{\langle l_1, 1\rangle} \in X^{2n}.$$

Consequently,

$$\sup_{f\in H^{\beta}_{\infty,R}(T)\cap X^{2n}} \|f\|_\infty \geq \|f_{x^*}\|_\infty$$

$$\geq \inf_{\substack{a\in R\\ \eta\in\Theta_{2n}(\varphi_0,\mathcal{K}_\beta)}} \|a + D_r * \varphi_0(\mathcal{K}_\beta * h_\eta)\|_\infty = \|\Phi^{\beta}_{n,r}\|_\infty.$$

Thus,

$$d^{2n}(H^{r,\beta}_{\infty,R}(T), L_\infty(T)) \geq \|\Phi^{\beta}_{n,r}\|_\infty.$$

The upper bound for $\lambda_{2n}(H^{\beta}_{\infty,R}(T), L_q(T))$ and $s_{2n}(H^{\beta}_{\infty,R}(T), L_q(T))$ follows from Theorem 2.18. From (4.2), (4.4), (4.3), and monotonicity of widths it follows that it remains to prove the upper bound for $\lambda_{2n-1}(H^{r,\beta}_{\infty,R}(T), L_\infty(T))$. By Theorem 1.6 there exists a linear optimal method of recovery of $f(0)$ on the class $H^{r,\beta}_{\infty,R}(T)$ which uses the information $a_0(f), \dots, a_{n-1}(f), b_1(f), \dots, b_{n-1}(f)$. Using Lemma 4.14, for such method we obtain

$$\sup_{f\in H^{r,\beta}_{\infty,R}(T)} |f(0) - c_0a_0(f) - \sum_{j=1}^{n-1}(c_ja_j(f) + d_jb_j(f))|$$

$$= \sup_{\substack{f\in H^{r,\beta}_{\infty,R}(T)\\ c_j(f)=0,\ |j|<n}} |f(0)| = \|\Phi^{\beta}_{n,r}\|_\infty. \quad (4.17)$$

Since $H^{r,\beta}_{\infty,R}(T)$ is a shift invariant class it follows from (4.17) that for all $t \in T$ and all $f \in H^{r,\beta}_{\infty,R}(T)$

$$|f(t) - c_0a_0(f) - \sum_{j=1}^{n-1}(\mu_j(t)a_j(f) + \nu_j(t)b_j(f))| \leq \|\Phi^{\beta}_{n,r}\|_\infty,$$

where

$$\mu_j(t) = c_j\cos jt - d_j\sin jt, \qquad \nu_j(t) = c_j\sin jt + d_j\cos jt.$$

Hence

$$\lambda_{2n-1}(H^{r,\beta}_{\infty,R}(T), L_\infty(T)) \leq \|\Phi^{\beta}_{n,r}\|_\infty.$$

□

It follows from this theorem and Theorem 2.18 that

$$i_{2n-1}(H^{\beta}_{\infty,R}(T), L_{\infty}(T)) < s_{2n-1}(H^{\beta}_{\infty,R}(T), L_{\infty}(T)).$$

In other words, if we have to choose $2n-1$ functionals, then sampling is worse than the Fourier coefficients. For the even case both of these types of information are optimal.

4.4 Estimations of n-Widths in Hilbert Spaces

In Theorem 4.3 the exact values of n-widths of H_p in the space $L_p(E,\mu)$, $1 \le q \le p \le \infty$, were found. For $1 \le p < q \le \infty$ even in the simple case when $E = T_\rho$ and $\mu = \sigma_\rho$ the exact result is known only for $p = 2$, $q = \infty$ for the linear and Gel'fand n-widths (Osipenko, Stessin [1990]).

In the case $p = 2$, $q = \infty$ there exists a general method of obtaining estimates for the linear and Gel'fand n-widths which is based on extremal properties of s-numbers of bounded linear operators. This method often gives exact results. In fact, the same method makes possible to obtain the Bernstein n-widths for $p = \infty$, $q = 2$.

Let H and H_1 be Hilbert spaces and $T \colon H \to H_1$ a bounded linear operator. Suppose that

$$T'T\varphi_k = s_k^2\varphi_k, \quad k = 1, 2, \dots, \tag{4.18}$$

where $s_1 \ge s_2 \ge \dots > 0$ and $\{\varphi_k\}$ form a complete orthonormal basis for the range of $T'T(H)$ (a sufficient condition is that T be a compact operator). The values $s_k(T) := s_k$ are called the *s-numbers* of T.

Set $\psi_k := s_k^{-1}(T)T\varphi_k$. It is easy to verify that $\{\psi_k\}$ is an orthonormal system in H_1. Then there exists the decomposition of T (see, for example, Gohberg, Krein [1965]) which is given by

$$T = \sum_{k=1}^{\infty} s_k(T)(\cdot, \varphi_k)\psi_k.$$

Lemma 4.16. *Let H and H_1 be Hilbert spaces, $T \colon H \to H_1$ a bounded linear operator and $\{\varphi_k\}_1^\infty$ form a complete orthonormal basis for the range of $T'T(H)$. If $\{T\varphi_k\}_1^\infty$ is an orthogonal system in H_1 and $\|T\varphi_1\|_{H_1} \ge \|T\varphi_2\|_{H_1} \ge \dots > 0$, then the condition (4.18) holds with $s_k = \|T\varphi_k\|_{H_1}$.*

Proof. Since $\{\varphi_k\}_1^\infty$ is an orthonormal basis in $T'T(H)$, for each $x \in H$ we have

$$x = \sum_{j=1}^{\infty}(x, \varphi_j)\varphi_j + y,$$

where $y \in \operatorname{Ker} T'T = \operatorname{Ker} T$. Thus

$$(T'T\varphi_k, x) = \left(T\varphi_k, \sum_{j=1}^{\infty}(x, \varphi_j)T\varphi_j\right) = \|T\varphi_k\|_{H_1}^2(\varphi_k, x).$$

Hence

$$T'T\varphi_k = \|T\varphi_k\|_{H_1}^2\varphi_k.$$

□

Theorem 4.17. *Assume that a bounded linear operator T satisfies the conditions* (4.18). *Then*

$$\sum_{k=1}^{n} s_k^2(T) = \max_{\{e_k\}_1^n} \sum_{k=1}^{n} \|Te_k\|_{H_1}^2, \tag{4.19}$$

where the maximum is taken over all orthonormal systems $\{e_k\}_1^n$ in H. Furthermore,

$$\sum_{k=n+1}^{\infty} s_k^2(T) = \min_{\{e_k\}_1^{\infty}} \sum_{k=1}^{\infty} \|Te_k\|_{H_1}^2, \tag{4.20}$$

where the minimum is taken over all orthonormal systems $\{e_k\}_1^{\infty}$ in H such that $\operatorname{codim} \operatorname{span}\{e_k\}_1^{\infty} = n$.

Proof. Let $\{e_k\}_1^n$ be any orthonormal system in H. Set $L_n := \operatorname{span}\{e_k\}_1^n$ and $T_n := T_{|L_n}$. Suppose that the Schmidt decomposition of T_n has the form

$$T_n = \sum_{k=1}^{n} s_k(T_n)(\cdot, \varphi_k')\psi_k'.$$

The value

$$\left(\sum_{k=1}^{n} \|T_n e_k\|_{H_1}^2\right)^{1/2}$$

is the Hilbert-Schmidt norm of T_n and does not depend on the choice of the orthonormal basis in L_n. Therefore,

$$\sum_{k=1}^{n} \|Te_k\|_{H_1}^2 = \sum_{k=1}^{n} \|T_n e_k\|_{H_1}^2 = \sum_{k=1}^{n} \|T_n\varphi_k'\|_{H_1}^2 = \sum_{k=1}^{n} s_k^2(T_n).$$

Denote by P_n an orthoprojector in H onto L_n. Using the properties of s-numbers of bounded linear operators (see Gohberg, Krein [1965, p. 82]) we have

$$s_k(T_n) = s_k(T \circ P_n) \le \|P_n\| s_k(T) = s_k(T). \tag{4.21}$$

Thus

$$\sum_{k=1}^{n} \|Te_k\|_{H_1}^2 \le \sum_{k=1}^{n} s_k^2(T).$$

If $e_k = \varphi_k$, $k = 1, \dots, n$, then

$$\sum_{k=1}^{n} \|T\varphi_k\|_{H_1}^2 = \sum_{k=1}^{n} s_k^2(T).$$

The equality (4.19) is proved.

Let $\operatorname{codim} \operatorname{span}\{e_k\}_1^\infty = n$ and $\{e_k\}_1^\infty$ be an orthonormal system in H. Then there exists an orthonormal system $\{e_k'\}_1^n$ in H such that $\{e_k'\}_1^n \cup \{e_k\}_1^\infty$ is an orthonormal basis. The value

$$\left(\sum_{k=1}^{n} \|Te_k'\|_{H_1}^2 + \sum_{k=1}^{\infty} \|Te_k\|_{H_1}^2 \right)^{1/2},$$

is the Hilbert-Schmidt norm of T and does not depend on the choice of the orthonormal basis in H. It is equal to

$$\left(\sum_{k=1}^{\infty} s_k^2(T) \right)^{1/2}.$$

Consequently, using (4.19), we obtain

$$\min_{\{e_k\}_1^\infty} \sum_{k=1}^{\infty} \|Te_k\|_{H_1}^2 = \sum_{k=1}^{\infty} s_k^2(T) - \max_{\{e_k'\}_1^n} \sum_{k=1}^{n} \|Te_k'\|_{H_1}^2 = \sum_{k=n+1}^{\infty} s_k^2(T).$$

□

Let H and X be normed linear spaces and $T\colon H \to X$ be a bounded linear operator. Set

$$\lambda_n(T) := \lambda_n(T(BH), X), \qquad d^n(T) := d^n(T(BH), X),$$
$$i_n(T) := i_n(T(BH), X),$$

where

$$BH := \{h \in H : \|h\|_H \le 1\}.$$

Proposition 4.18. *If H is a Hilbert space and X_r is an r-dimensional subspace of X, then*

$$\lambda_n(T(BH)+X_r,X)=d^n(T(BH)+X_r,X)=i_n(T(BH)+X_r,X).$$

In particular, for $r=0$

$$\lambda_n(T)=d^n(T)=i_n(T).$$

Proof. In view of Proposition 4.4 it remains to prove the inequality

$$\lambda_n(T(BH)+X_r,X)\le d^n(T(BH)+X_r,X). \tag{4.22}$$

Let Y be a normed linear space, $T(BH)+X_r\in Y$, $z'_1,\dots,z'_{n+r}\in Y^*$, and

$$Y^{n+r}:=\{\,y\in Y\mid \langle z'_j,y\rangle=0,\ j=1,\dots,n+r\,\}.$$

Assume that

$$\sup_{x\in(T(BH)+X_r)\cap Y^{n+r}}\|x\|_X<\infty.$$

Then $X_r\cap Y^{n+r}=0$. Consequently, there exists a basis $y'_1,\dots,y'_{n+r}$ in $\operatorname{span}\{z'_j\}_1^{n+r}$ such that $y'_1,\dots,y'_r$ are linearly independent on X_r and for all $x\in X_r$

$$\langle y'_{r+1},x\rangle=\dots=\langle y'_{n+r},x\rangle=0.$$

Denote by $x_1,\dots,x_r\in X_r$ a biorthogonal system to $y'_1,\dots,y'_r$. For all $k=1,\dots,r$ and $j=1,\dots,n+r$ we have

$$\langle y'_j,x_k\rangle=\begin{cases}1, & j=k,\\ 0, & j\ne k.\end{cases} \tag{4.23}$$

Denote by $\{e_j\}_1^m$, $m\le n$, an orthonormal basis in $\operatorname{span}\{T'y'_{r+1},\dots,T'y'_{n+r}\}$ (here by T' we denote the adjoint operator of T, where T is considered as an operator from H into Y). Suppose that

$$e_j=\sum_{k=r+1}^{n+r}\alpha_{jk}T'y'_k,\quad j=1,\dots,m.$$

Set

$$u'_j:=\sum_{k=r+1}^{n+r}\alpha_{jk}y'_k,\quad P_0y:=\sum_{j=1}^{m}\langle u'_j,y\rangle Te_j,\quad Py:=P_0y+\sum_{j=1}^{r}\langle y'_j,y-P_0y\rangle x_j.$$

In view of (4.23) for all $u \in X_r$, $P_0 u = 0$ and $Pu = u$. Moreover,

$$P_0(Th) = \sum_{j=1}^{m} \langle u_j', Th\rangle Te_j = \sum_{j=1}^{m} (h, T'u_j')Te_j = \sum_{j=1}^{m} (h, e_j)Te_j.$$

Putting $z := h - \sum_{j=1}^{m}(h, e_j)e_j$, we obtain

$$Th - P_0(Th) = Tz.$$

If $h \in BH$, then $\|z\|_H \le \|h\|_H \le 1$. Thus for all $h \in BH$ and $u \in X_r$

$$\begin{aligned} Th + u - P(Th + u) &= Th - P(Th) \\ &= Th - P_o(Th) - \sum_{j=1}^{r} \langle y_j', Th - P_0(Th)\rangle x_j = Tz - \sum_{j=1}^{r} \langle y_j', Tz\rangle x_j, \end{aligned}$$

where $z \in BH$. Since for $k = r+1, \dots, n+r$

$$\langle y_k', Tz\rangle = (z, T'y_k') = 0,$$

we have

$$Tz - \sum_{j=1}^{r} \langle y_j', Tz\rangle x_j \in (T(BH) + X_r) \cap Y^{n+r}.$$

Therefore,

$$\begin{aligned} \lambda_{n+r}(T(BH) + X_r, X) &\le \sup_{\substack{h \in BH \\ u \in X_r}} \|Th + u - P(Th + u)\|_X \\ &\le \sup_{x \in (T(BH) + X_r) \cap Y^{n+r}} \|x\|_X. \end{aligned}$$

Hence we obtain (4.22). □

To obtain estimates of n-widths we need the following simple result.

Proposition 4.19. *Let H be a Hilbert space, $\omega := \dim H$, and $T \colon H \to C(E)$ a bounded linear operator. Then*

$$\|T\|_{H \to C(E)} = \sup_{z \in E} \left(\sum_{k=1}^{\omega} |(Te_k)(z)|^2 \right)^{1/2}$$

for any orthonormal basis $\{e_k\}_1^{\omega}$ in H.

Proof. We have

$$\|T\|_{H\to C(E)} = \sup_{h\in BH}\sup_{z\in E}|(Th)(z)| = \sup_{z\in E}\sup_{h\in BH}|(Th)(z)|$$
$$= \sup_{z\in E}\sup_{\{c_k\}_1^\omega\in Bl_2^\omega}\left|\sum_{k=1}^{\omega} c_k(Te_k)(z)\right| = \sup_{z\in E}\left(\sum_{k=1}^{\omega}|(Te_k)(z)|^2\right)^{1/2},$$

where

$$Bl_2^\omega = \left\{ (c_1, c_2, \dots) \,\middle|\, \sum_{k=1}^{\omega}|c_k|^2 \le 1 \right\}.$$

□

Theorem 4.20. *Let H be a Hilbert space, E a topological space with probability measure ν such that* $\operatorname{supp}\nu = E$, *and $T\colon H\to C(E)$ a bounded linear operator. Define $T_0\colon H\to L_2(E,\nu)$ by the equation $T_0h := Th$. Assume that T_0 satisfies the conditions* (4.18). *Then*

$$\left(\sum_{k=n+1}^{\infty} s_k^2\right)^{1/2} \le \lambda_n(T) = d^n(T) = i_n(T) \le \sup_{z\in E}\left(\sum_{k=n+1}^{\infty}|(T\varphi_k)(z)|^2\right)^{1/2}.$$

Proof. Since $\operatorname{supp}\nu = E$, $\operatorname{Ker}T_0'T_0 = \operatorname{Ker}T_0 = \operatorname{Ker}T$ and we can assume, without loss of generality, that $\{\varphi_k\}$ is an orthonormal basis in H. Let H^n be an arbitrary n-codimensional subspace of H. Assume that $\{\varphi_k'\}$ is an orthonormal basis of H^n. We obtain from Proposition 4.19 and (4.20)

$$\|T\|_{H^n\to C(E)}^2 = \sup_{z\in E}\sum_{k=1}^{\infty}|(T\varphi_k')(z)|^2 \ge \int_E\sum_{k=1}^{\infty}|(T\varphi_k')(z)|^2\,d\nu(z)$$
$$= \sum_{k=1}^{\infty}\|T_0\varphi_k'\|_{L_2(E,\nu)}^2 \ge \sum_{k=n+1}^{\infty} s_k^2. \quad (4.24)$$

Let Y be an arbitrary normed linear space containing $T(BH)$, Y^n an n-codimensional subspace of Y, and $y_1',\dots,y_n'\in Y^*$ such that

$$Y^n = \{\, y\in Y : \langle y_k', y\rangle = 0,\ k=1,\dots,n\,\}.$$

Set $e_k := T'y_k'$ (here by T' we mean the adjoint operator of T where the last is considered as an operator from H into Y). In view of (4.24) we have

$$\sup_{\substack{h\in BH\\ \langle y_k', Th\rangle=0,\ k=1,\dots,n}}\|Th\|_{C(E)} = \sup_{\substack{h\in BH\\ (h,e_k)=0,\ k=1,\dots,n}}\|Th\|_{C(E)} \ge \sum_{k=n+1}^{\infty} s_k^2.$$

Since Y and Y^n are arbitrary, we obtain

$$d^n(T) \geq \sum_{k=n+1}^{\infty} s_k^2.$$

To prove the upper bound we set $\psi_k := T\varphi_k$ and consider the linear operator $P\colon C(E) \to C(E)$ defined by

$$Py := \sum_{k=1}^{n} (y, \psi_k)_{L_2(E,\nu)} s_k^{-2} \psi_k.$$

For $y = Th$ we have

$$P(Th) = \sum_{k=1}^{n} (Th, \psi_k)_{L_2(E,\nu)} s_k^{-2} \psi_k = \sum_{k=1}^{n} (h, \varphi_k) T\varphi_k.$$

Thus,

$$\begin{aligned}\lambda_n(T) &\leq \sup_{h\in BH} \|Th - P(Th)\|_{C(E)} = \sup_{h\in BH} \left\| T\left(h - \sum_{k=1}^{n} (h,\varphi_k)\varphi_k \right) \right\|_{C(E)} \\ &= \sup_{h\in BH} \left\| \sum_{k=n+1}^{\infty} (h,\varphi_k)\psi_k \right\|_{C(E)} \leq \sup_{z\in E} \left(\sum_{k=n+1}^{\infty} |\psi_k(z)|^2 \right)^{-1/2}\end{aligned}$$

□

Theorem 4.21. *Let H, E, ν, and T_0 be as above. Assume that X_r is an r-dimensional subspace of $C(E)$ such that $X_r \perp T_0 H$ in $L_2(E, \nu)$. Then*

$$\begin{aligned}\left(\sum_{k=n+1}^{\infty} s_k^2 \right)^{1/2} &\leq \lambda_{n+r}(T(BH) + X_r, C(E)) = d^{n+r}(T(BH) + X_r, C(E)) \\ &= i_{n+r}(T(BH) + X_r, C(E)) \leq \sup_{z\in E} \left(\sum_{k=n+1}^{\infty} |(T\varphi_k)(z)|^2 \right)^{1/2}.\end{aligned}$$

Proof. Let $e_1, \dots, e_r$ be an orthonormal basis of X_r in $L_2(E, \nu)$. Denote by $H_{r,\varepsilon}$, $\varepsilon > 0$, the Hilbert space of elements $\{f, g\}$, $f \in H$, $g \in X_r$, with the inner product

$$(\{f_1, g_1\}, \{f_2, g_2\}) := (f_1, g_1) + \varepsilon \sum_{j=1}^{r} c_j \bar{d}_j,$$

where

$$g_1 = \sum_{j=1}^{r} c_j e_j, \qquad g_2 = \sum_{j=1}^{r} d_j e_j.$$

Put $L\{f, g\} := Tf + g$. Denote by L_0 the operator L as an operator from $H_{r,\varepsilon}$ into $L_2(E, \nu)$. Then

$$L_0' L_0 \{f, g\} = \{T_0' T_0 f, \varepsilon^{-1} g\}.$$

Set

$$\phi_j := \{0, \varepsilon^{-1/2} e_j\}, \quad j = 1, \dots, r, \qquad \phi_j := \{\varphi_{j-r}, 0\}, \quad j = r+1, \dots .$$

The elements $\phi_1, \phi_2, \dots$ form a complete orthonormal basis for the range of $L_0' L_0$ and

$$L_0' L_0 \phi_j = \varepsilon^{-1} \phi_j, \quad j = 1, \dots, r, \qquad L_0' L_0 \phi_j = s_{j-r}^2 \phi_j, \quad j = r+1, \dots .$$

From Theorem 4.20 for $\varepsilon \le s_1^{-2}$ we have

$$d^{n+r}(L(BH_{r,\varepsilon}), C(E)) \ge \left(\sum_{k=n+1}^{\infty} s_k^2 \right)^{1/2}.$$

Since $T(BH) + X_r \supset L(BH_{r,\varepsilon})$,

$$d^{n+r}(T(BH) + X_r, C(E)) \ge d^{n+r}(L(BH_{r,\varepsilon}), C(E)) \ge \left(\sum_{k=n+1}^{\infty} s_k^2 \right)^{1/2}.$$

It follows from Proposition 4.18 that

$$\begin{aligned} \lambda_{n+r}(T(BH) + X_r, C(E)) &= d^{n+r}(T(BH) + X_r, C(E)) \\ &= i_{n+r}(T(BH) + X_r, C(E)). \end{aligned}$$

The upper bound follows from the obvious inequality

$$\lambda_{n+r}(T(BH) + X_r, C(E)) \le \lambda_n(T(BH), C(E))$$

and Theorem 4.20. □

Now we will obtain similar estimates for the Bernstein n-widths. Let W be a closed, convex, centrally symmetric subset of a normed linear space Y. The *Bernstein* n-*width* of W is defined by

$$b_n(W, Y) := \sup_{Y_{n+1}} \sup_{\substack{r>0 \\ rBY_{n+1} \subset W}} r,$$

where Y_{n+1} is any $(n+1)$-dimensional subspace of Y. Let X and Y be normed linear spaces. For a bounded linear operator $T\colon X \to Y$ set

$$b_n(T) := b_n(T(BX), Y).$$

If $\operatorname{Ker} T = 0$ and $Y_{n+1} \subset T(X)$, then it is easily shown that

$$\sup_{\substack{r>0 \\ rBY_{n+1} \subset T(BX)}} r = \inf_{\substack{Tx \in Y_{n+1} \\ x \neq 0}} \frac{\|Tx\|_Y}{\|x\|_X}.$$

Thus for $\operatorname{Ker} T = 0$

$$b_n(T) = \sup_{Y_{n+1} \subset T(X)} \inf_{\substack{Tx \in Y_{n+1} \\ x \neq 0}} \frac{\|Tx\|_Y}{\|x\|_X}.$$

Theorem 4.22. *Let H be a Hilbert space, E a topological space with probability measure ν such that* $\operatorname{supp} \nu = E$, *$H_1$ a subspace of $L_2(E, \nu)$, and $X_E \subset H_1$ a subspace of $C(E)$. Suppose that a bounded linear operator $T_0\colon H_1 \to H$ satisfies the conditions* (4.18) *and $\varphi_k \in X_E$, $k = 1, 2, \dots$. Set $T := T_0|_{X_E}$. Then*

$$\left(\sup_{z \in E} \sum_{k=1}^{n+1} s_k^{-2} |\varphi_k(z)|^2\right)^{-1/2} \le b_n(T) \le \left(\sum_{k=1}^{n+1} s_k^{-2}\right)^{-1/2}.$$

Proof. Without loss of generality we can assume that φ_k form an orthonormal basis in H_1. Then $\operatorname{Ker} T = 0$. Let $Y_{n+1} \subset T(X)$ and $X_{n+1} = T^{-1}(Y_{n+1})$. Set $T_{n+1} := T|_{X_{n+1}}$. Assume that the Schmidt decomposition of T_{n+1} has the form

$$T_{n+1} = \sum_{k=1}^{n} s_k(T_{n+1})(\cdot, \varphi_k')\psi_k'.$$

Then from Proposition 4.19 we have

$$\inf_{\substack{Tf \in Y_{n+1} \\ f \neq 0}} \frac{\|Tf\|_H}{\|f\|_{C(E)}} = \inf_{\substack{g \in Y_{n+1} \\ g \neq 0}} \frac{\|g\|_H}{\|T_{n+1}^{-1} g\|_{C(E)}} = \|T_{n+1}^{-1}\|_{Y_{n+1} \to C(E)}^{-1}$$

$$= \left(\sup_{z \in E} \sum_{k=1}^{n+1} |(T_{n+1}^{-1}\psi_k')(z)|^2\right)^{-1/2} \le \left(\sum_{k=1}^{n+1} \int_E |(T_{n+1}^{-1}\psi_k')(z)|^2 \, d\nu(z)\right)^{-1/2}$$

$$= \left(\sum_{k=1}^{n+1} \|T_{n+1}^{-1}\psi_k'\|_{H_1}^2\right)^{-1/2} = \left(\sum_{k=1}^{n+1} s_k^{-2}(T_{n+1})\right)^{-1/2}.$$

Since $s_k(T_{n+1}) \le s_k(T_0) = s_k$, $k = 1, \dots, n+1$, we obtain

$$b_n(T) \le \left(\sum_{k=1}^{n+1} s_k^{-2}\right)^{-1/2}.$$

Let $Y_{n+1} = \operatorname{span}\{T\varphi_k\}_1^{n+1}$. Then $\psi_k := s_k^{-1} T\varphi_k$, $k = 1, \dots, n+1$, form an orthonormal basis in Y_{n+1}. Thus,

$$b_n(T) \ge \inf_{\substack{Tf \in Y_{n+1} \\ f \ne 0}} \frac{\|Tf\|_H}{\|f\|_{C(E)}} = \left(\sup_{z \in E} \sum_{k=1}^{n+1} |(T_{n+1}^{-1}\psi_k)(z)|^2\right)^{-1/2}$$

$$= \left(\sup_{z \in E} \sum_{k=1}^{n+1} s_k^{-2} |\varphi_k(z)|^2\right)^{-1/2}.$$

□

4.5 Diagonal Operators

Consider the operator $T\colon l_2 \to l_\infty$ defined by the equation

$$T(\{x_k\}_1^\infty) := \{a_k x_k\}_1^\infty, \tag{4.25}$$

where $a_1 \ge a_2 \ge \dots > 0$ and $\lim_{k\to\infty} a_k = 0$. We shall be interested in estimates of n-widths of T. In other words, we would like to calculate n-widths of ellipsoid

$$\sum_{k=1}^{\infty} \frac{x_k^2}{a_k^2} \le 1$$

in l_∞.

Theorem 4.23.

$$\lambda_n(T) = d^n(T) = i_n(T) = \sup_{m>n} \left(\frac{m-n}{\sum_{k=1}^m a_k^{-2}}\right)^{1/2}.$$

To prove this theorem we need some preliminary results.

Lemma 4.24. *Let $0 \le A_j \le 1$, $j = 1, \dots, r$, and $\sum_{j=1}^r A_j = n$. Then there exist n orthonormal vectors $f_k \in R^r$, $k = 1, \dots, n$, for which*

$$\sum_{k=1}^n (f_k)_j^2 = A_j, \quad j = 1, \dots, r.$$

Proof. The proof is by induction on r. For $r = n$ the statement is evident since it suffices to take the standard basis

$$(f_k)_j = \begin{cases} 1, & j = k, \\ 0, & j \neq k. \end{cases}$$

Suppose that $r > n$ and the statement is valid for $r-1$. Without loss of generality we can assume that $A_r = \min_{1 \le j \le r} A_j$. Since

$$\sum_{j=1}^{r-1} (1 - A_j) = r - 1 + A_r - n \ge A_r,$$

there exist B_j, $j = 1, \dots, r-1$, which satisfy the conditions

$$A_j \le B_j \le 1, \qquad \sum_{j=1}^{r-1} B_j = \sum_{j=1}^{r} A_j = n.$$

Put $B_r := 0$. By the induction hypothesis there exist orthonormal vectors $g_k \in R^r$, $k = 1, \dots, n$, such that

$$\sum_{k=1}^{n} (g_k)_j^2 = B_j, \quad j = 1, \dots, r-1.$$

For $1 \le s \le r-1$ consider the transformation of R^r defined by

$$(U_s(t)x)_j := \begin{cases} x_j, & j \neq s, r, \\ x_s \cos t + x_r \sin t, & j = s, \\ -x_s \sin t + x_r \cos t, & j = r. \end{cases}$$

It is easy to see that for all $t \in T$ the transformation $U_s(t)$ is unitary. Put

$$B_{js}(t) := \sum_{k=1}^{n} (U_s(t) g_k)_j^2.$$

Then for all $t \in T$, $0 \le B_{js}(t) \le 1$, $j = 1, \dots, r$, $\sum_{j=1}^{r} B_{js}(t) = n$, and $B_{js}(t) = B_j$, $j \neq s, r$. Furthermore, $B_{ss}(0) = B_s$ and $B_{ss}(\pi/2) = B_r$. Consequently, there exists a $t_s \in [0, \pi/2]$ for which $B_{ss}(t_s) = A_s$. Using sequentially the transformation $U_s(t_s)$ for $s = 1, \dots, j$, we obtain the system of numbers $(A_1, \dots, A_j, B_{j+1}, \dots, B_{r-1}, B_r^{(j)})$ (and the corresponding orthonormal system of vectors). The inequalities $B_r^{(j)} \le A_r \le A_{j+1} \le B_{j+1}$ give the possibility to make every next step, that is, to find t_{j+1}. For $j = r-1$ we obtain the required orthonormal system. □

Lemma 4.25. *Let $a_1 \geq a_2 \geq \ldots > 0$, $\lim_{k\to\infty} a_k = 0$, and $r > n$ such that*

$$\frac{r-n}{\sum_{k=1}^{r} a_k^{-2}} = \sup_{m>n} \frac{m-n}{\sum_{k=1}^{m} a_k^{-2}}. \tag{4.26}$$

Then

$$a_{r+1}^2 \leq \frac{r-n}{\sum_{k=1}^{r} a_k^{-2}} \leq a_r^2. \tag{4.27}$$

Proof. Put $s := [m/2]$. For $m > n$

$$\frac{m-n}{\sum_{k=1}^{m} a_k^{-2}} \leq \frac{m-n}{s a_1^{-2} + s a_s^{-2}} \to 0 \quad \text{as } m \to \infty.$$

From here it follows the existence of r for which the upper bound in (4.26) is attained.

By virtue of definition of r we have

$$\frac{r+1-n}{\sum_{k=1}^{r+1} a_k^{-2}} \leq \frac{r-n}{\sum_{k=1}^{r} a_k^{-2}}.$$

Thus,

$$(r+1-n)\sum_{k=1}^{r} a_k^{-2} \leq (r-n)\left(\sum_{k=1}^{r} a_k^{-2} + a_{r+1}^{-2}\right).$$

Consequently,

$$a_{r+1}^2 \leq \frac{r-n}{\sum_{k=1}^{r} a_k^{-2}}.$$

We prove now that

$$\frac{r-n}{\sum_{k=1}^{r} a_k^{-2}} \leq a_r^2. \tag{4.28}$$

For $r = n+1$ this inequality is obvious. Assume that $r > n+1$. Then

$$\frac{r-1-n}{\sum_{k=1}^{r-1} a_k^{-2}} \leq \frac{r-n}{\sum_{k=1}^{r} a_k^{-2}}.$$

Thus,

$$(r-1-n)\sum_{k=1}^{r} a_k^{-2} \leq (r-n)\left(\sum_{k=1}^{r} a_k^{-2} - a_r^{-2}\right).$$

From here it is easily follows (4.28). □

Proof of Theorem 4.23. Let $0 < \varepsilon < 1$ and r satisfies the condition of Lemma 4.25. Define the probability measure on N by

$$\nu(\{j\}) := \begin{cases} (1-\varepsilon)\dfrac{a_j^{-2}}{\sum_{k=1}^{r} a_k^{-2}}, & 1 \le j \le r, \\ \dfrac{\varepsilon}{2^{j-r}}, & j > r. \end{cases}$$

Denote by T_0 the operator T which is considered as an operator from l_2 into $l_2(\nu)$. Then for the standard basis e_j we have

$$T_0' T_0 e_j = s_j^2 e_j, \quad j = 1, 2, \dots, \tag{4.29}$$

where

$$s_j^2 = \begin{cases} \dfrac{1-\varepsilon}{\sum_{k=1}^{r} a_k^{-2}}, & 1 \le j \le r, \\ \dfrac{\varepsilon a_j^2}{2^{j-r}}, & j > r. \end{cases}$$

Put

$$A_j := (r-n)\frac{a_j^{-2}}{\sum_{k=1}^{r} a_k^{-2}}, \quad 1 \le j \le r.$$

By Lemma 4.25 it follows that $A_1 \le \dots \le A_r \le 1$. Furthermore,

$$\sum_{j=1}^{r} A_j = r - n.$$

Thus using Lemma 4.24 we can obtain an orthonormal system $f_{n+1}, \dots, f_r \in R^r$ for which

$$\sum_{k=n+1}^{r} (f_k)_j^2 = A_j, \quad 1 \le j \le r.$$

Let vectors $f_1, \dots, f_n$ complete this system to an orthonormal basis in R^r. Consider the orthonormal basis in l_2

$$g_j := \begin{cases} f_j, & 1 \le j \le r, \\ e_j, & j > r. \end{cases}$$

Since vectors f_j, $1 \le j \le r$, can be represented as linear combinations of $e_1, \dots, e_r$, the equalities (4.29) remain valid for the basis $\{g_j\}$. By Theorem 4.20 for sufficiently small ε we have

$$\left(\sum_{k=n+1}^{\infty} s_k^2\right)^{1/2} \le \lambda_n(T) = d^n(T) = i_n(T) \le \sup_{j\in N}\left(\sum_{k=n+1}^{\infty} (Tg_k)_j^2\right)^{1/2}$$
$$= \left(\max\left\{\frac{r-n}{\sum_{k=1}^{r} a_k^{-2}}, a_{r+1}^2, a_{r+2}^2, \dots\right\}\right)^{1/2} = \left(\frac{r-n}{\sum_{k=1}^{r} a_k^{-2}}\right)^{1/2}$$

(the last equality follows in view of (4.27)). Since

$$\sum_{k=n+1}^{\infty} s_k^2 = (1-\varepsilon)\frac{r-n}{\sum_{k=1}^{r} a_k^{-2}} + \varepsilon \sum_{j=r+1}^{\infty} \frac{a_j^2}{2^{j-r}} \to \frac{r-n}{\sum_{k=1}^{r} a_k^{-2}} \quad \text{as } \varepsilon \to 0,$$

passing ε to zero, we obtain the statement of theorem. □

By the analogy with Lemma 4.25 the following result can be proved.

Lemma 4.26. *Let $a_1 \ge \dots \ge a_N > 0$, $N > n$, and $0 \le r \le n$ such that*

$$\frac{n-r+1}{\sum_{k=r+1}^{N} a_k^2} = \max_{0\le m\le n} \frac{n-m+1}{\sum_{k=m+1}^{N} a_k^2}.$$

Then

$$a_r^{-2} \le \frac{n-r+1}{\sum_{k=r+1}^{N} a_k^2} \le a_{r+1}^{-2} \qquad (a_0^{-2} := 0). \tag{4.30}$$

A result similar to Theorem 4.23 can be obtained for the Bernstein n-widths.

Theorem 4.27. *Let $\{a_k\}_1^\infty \in l_2$. Assume that the operator $T\colon l_\infty \to l_2$ is defined by* (4.25). *Then*

$$b_n(T) = \min_{0\le m\le n} \left(\frac{\sum_{k=m+1}^{\infty} a_k^2}{n-m+1}\right)^{1/2}.$$

Proof. 1. The upper bound. Let $0 < \varepsilon < 1$ and $0 \le m \le n$. Define the probability measure on N by

$$\nu_m(\{j\}) := \begin{cases} (1-\varepsilon)\dfrac{a_j^2}{\sum_{k=m+1}^{\infty} a_k^2}, & j > m, \\ \dfrac{\varepsilon}{m}, & j \le m. \end{cases}$$

Put $T_0 := T_{|l_2(\nu_m)}$. It is easily seen that

$$T_0' T_0 e_j = s_j^2 e_j, \quad j = 1, 2, \dots ,$$

where

$$s_j^2 = \begin{cases} (1-\varepsilon)^{-1} \sum\limits_{k=m+1}^{\infty} a_k^2, & j > m, \\ \varepsilon^{-1} a_j^2 m, & j \le m. \end{cases}$$

By Theorem 4.22 for sufficiently small ε we have

$$b_n(T) \le \left((1-\varepsilon) \frac{n-m+1}{\sum_{k=m+1}^{\infty} a_k^2} + \frac{\varepsilon}{m} \sum_{j=1}^{m} a_j^{-2} \right)^{-1/2} .$$

Passing to the limit as $\varepsilon \to 0$, we obtain

$$b_n(T) \le \min_{0 \le m \le n} \left(\frac{\sum_{k=m+1}^{\infty} a_k^2}{n-m+1} \right)^{1/2} .$$

2. The lower bound. Let $N > n$ and $T_N := T_{|l_2^N}$. Define the probability measure on the set $\{1, \dots , N\}$ by

$$\mu(\{j\}) := \frac{a_j^2}{\sum_{k=1}^{N} a_k^2}.$$

Then for the operator $T_{N0} := T_{|l_2^N(\mu)}$ we have

$$T_{N0}' T_{N0} e_j = s^2 e_j, \quad j = 1, \dots , N,$$

where $\{e_j\}$ is the standard basis in R^N and

$$s^2 = \sum_{k=1}^{N} a_k^2.$$

Assume that r satisfies the conditions of Lemma 4.26. Put

$$A_j := (n-r+1) \frac{a_j^2}{\sum_{k=r+1}^{N} a_k^2}, \quad j = r+1, \dots , N.$$

It follows from Lemma 4.26 that $A_N \le \ldots \le A_{r+1} \le 1$. Moreover,

$$\sum_{j=r+1}^{N} A_j = n - r + 1.$$

In view of Lemma 4.24 there exists an orthonormal system of vectors $f_{r+1}, \ldots, f_{n+1} \in l_2^{N-r}$ for which

$$\sum_{k=r+1}^{n+1} (f_k)_j^2 = A_j, \quad j = r+1, \ldots, N.$$

Define vectors $g_1, \ldots, g_{n+1} \in l_2^N$ by

$$g_k := \frac{s}{a_k} e_k, \qquad k = 1, \ldots, r,$$

$$(g_k)_j := \begin{cases} 0, & 1 \le j \le r, \\ \dfrac{s}{a_j}(f_k)_j, & r < j \le N, \end{cases} \qquad k = r+1, \ldots, n+1.$$

It is easy to check that $\{g_k\}_1^N$ is an orthonormal basis in $l_2^N(\mu)$. Thus from Theorem 4.22 we have

$$b_n(T) \ge b_n(T_N) \ge \left(\max_{1 \le j \le N} \sum_{k=1}^{n+1} s^{-2} (g_k)_j^2 \right)^{-1/2}$$

$$= \left(\max \left\{ a_1^{-2}, \ldots, a_r^{-2}, \frac{n-r+1}{\sum_{k=r+1}^N a_k^2} \right\} \right)^{-1/2} = \left(\frac{n-r+1}{\sum_{k=r+1}^N a_k^2} \right)^{-1/2}$$

(the last equality follows from (4.30)). Consequently, for all $N > n$

$$b_n(T) \ge \max_{0 \le m \le n} \left(\frac{n-m+1}{\sum_{k=m+1}^N a_k^2} \right)^{-1/2}.$$

Passing $N \to \infty$ we obtain the required estimate. □

4.6 Hardy-Sobolev and Bergman-Sobolev Classes

Denote by $\mathcal{H}_\gamma$ the space of holomorphic functions in the unit disk D

$$f(z) = \sum_{k=0}^{\infty} a_k z^k \tag{4.31}$$

which satisfy the condition

$$\sum_{k=0}^{\infty} \gamma_k |a_k|^2 < \infty, \tag{4.32}$$

where γ_k is a sequence of non-negative numbers such that

$$\liminf_{k\to\infty} \gamma_k^{1/k} \geq 1$$

(the last condition implies that the function (4.31) is holomorphic in D if (4.32) is valid). Denote by H_γ the class of functions from $\mathcal{H}_\gamma$ for which

$$\sum_{k=0}^{\infty} \gamma_k |a_k|^2 \leq 1.$$

Theorem 4.28. *Let $\gamma_0 = \ldots = \gamma_{r-1} = 0$ and $\gamma_k > 0$ for $k \geq r$. Assume that $n \geq r$, $0 < \rho < 1$, and for all $k \geq n$*

$$\gamma_k^{-1} \rho^{2k} \leq \min_{m<n} \gamma_m^{-1} \rho^{2m}. \tag{4.33}$$

Then

$$\lambda_n(H_\gamma, C(T_\rho)) = d^n(H_\gamma, C(T_\rho)) = i_n(H_\gamma, C(T_\rho)) = \left(\sum_{k=n}^{\infty} \gamma_k^{-1} \rho^{2k}\right)^{1/2},$$

where $T_\rho := \{z \in C : |z| = \rho\}$.

Proof. Consider the Hilbert space

$$\mathcal{H}_\gamma^0 := \{\, f \in \mathcal{H}_\gamma : f^{(j)}(0) = 0,\ j = 0, \ldots, r-1 \,\}$$

with the inner product

$$(f, g) = \sum_{k=r}^{\infty} \gamma_k a_k \bar{b}_k,$$

where

$$f(z) = \sum_{k=r}^{\infty} a_k z^k, \qquad g(z) = \sum_{k=r}^{\infty} b_k z^k.$$

Define the probability measure on T_ρ by

$$d\sigma_\rho(\rho e^{i\theta}) = \frac{1}{2\pi}\, d\theta.$$

The functions

$$\varphi_k(z) := \gamma_k^{-1/2} z^k, \quad k = r, r+1, \dots ,$$

is an orthonormal basis in $\mathcal{H}_\gamma^0$. They form an orthonormal system in $L_2(T_\rho, \sigma_\rho)$ and

$$\|\varphi_k\|^2_{L_2(T_\rho,\sigma_\rho)} = \gamma_k^{-1} \rho^{2k}.$$

Since

$$H_\gamma = B\mathcal{H}_\gamma^0 + \mathcal{P}_r,$$

where $\mathcal{P}_r$ is the space of polynomials of degree at least $r-1$, the application of Lemma 4.16 and Theorem 4.21 gives the required statement. □

Now we consider some examples of the spaces H_γ.

The *Hardy–Sobolev class* H_2^r is the set of all functions analytic in the unit disk D for which $f^{(r)} \in H_2$. In this case $H_2^r = H_\gamma$ for

$$\gamma_0 = \dots = \gamma_{r-1} = 0, \qquad \gamma_k = \left(\frac{k!}{(k-r)!}\right)^2, \quad k \geq r.$$

The *Bergman class* A_2 is the set of all functions analytic in D for which

$$\frac{1}{\pi} \int_D |f(z)|^2 \, d\nu(z) \leq 1,$$

where ν is the normalized Lebesgue measure. The *Bergman–Sobolev class* A_2^r is the set of all functions analytic in D for which $f^{(r)} \in A_2$. The class A_2^1 is usually called the *Dirichlet class*. For A_2^r we have $A_2^r = H_\gamma$ for

$$\gamma_0 = \dots = \gamma_{r-1} = 0, \qquad \gamma_k = \frac{(k!)^2}{(k-r)!(k-r+1)!}, \quad k \geq r.$$

From Theorem 4.28 we obtain

Corollary 4.29. 1. *For all* $0 < \rho < 1$, $r \geq 0$, *and* $n \geq r$

$$\lambda_n(H_2^r, C(T_\rho)) = d^n(H_2^r, C(T_\rho)) = i_n(H_2^r, C(T_\rho)) = \left(\sum_{k=n}^{\infty} \left(\frac{(k-r)!}{k!}\right)^2 \rho^{2k}\right)^{1/2}.$$

2. *For all* $0 < \rho < 1$, $r \geq 1$, *and* $n \geq r$

$$\lambda_n(A_2^r, C(T_\rho)) = d^n(A_2^r, C(T_\rho)) = i_n(A_2^r, C(T_\rho)) = \left(\sum_{k=n}^{\infty} \frac{(k-r)!(k-r+1)!}{(k!)^2} \rho^{2k}\right)^{1/2}.$$

3. *For all $n \in N$ and $0 < \rho \leq (n+1)^{-\frac{1}{2n}}$*

$$\lambda_n(A_2, C(T_\rho)) = d^n(A_2, C(T_\rho)) = i_n(A_2, C(T_\rho)) = \frac{\rho^n}{1-\rho^2}(1 + n(1-\rho^2))^{1/2}.$$

In the case of A_2 we did not obtain the values of n-widths for all ρ. It is connected with the fact that the sequence

$$\gamma_k^{-1}\rho^{2k} = (k+1)\rho^{2k}$$

is not decreasing. Nevertheless, for $(n+1)\rho^{2n} \leq 1$ the condition (4.33) holds.

In a general case the corresponding result can be also obtained for all $0 < \rho < 1$:

$$\lambda_n(A_2, C(T_\rho)) = d^n(A_2, C(T_\rho)) = i_n(A_2, C(T_\rho)) = \left(\sum_{k=n}^{\infty} a_k\right)^{1/2},$$

where $\{a_k\}_0^\infty$ is the sequence $\{(k+1)\rho^{2k}\}_0^\infty$ ordered in non-increasing order.

The periodic analogue of $\mathcal{H}_\gamma$ is the space $\mathcal{H}_\gamma(T)$ which is the set of 2π-periodic functions analytic in T

$$f(z) = \sum_{k=-\infty}^{+\infty} a_k e^{ikz},$$

satisfying the condition

$$\sum_{k=-\infty}^{+\infty} \gamma_k |a_k|^2 < \infty,$$

where γ_k is a sequence of nonnegative numbers such that

$$\liminf_{k \to \pm\infty} \gamma_k^{1/|k|} > 1.$$

Denote by $H_\gamma(T)$ the class of functions from $\mathcal{H}_\gamma(T)$ for which

$$\sum_{k=-\infty}^{+\infty} \gamma_k |a_k|^2 \leq 1.$$

Similar to Theorem 4.28 the following result can be proved.

Theorem 4.30. *Assume that* $\gamma_0 = \ldots = \gamma_{r-1} = 0$, $\gamma_{r+1}^{-1} \geq \gamma_{r+2}^{-1} \cdots > 0$, *and* $\gamma_k = \gamma_{-k}$ *for all* $k \in N$. *Then for all* $n \geq r$ *and* $0 < \rho < 1$

$$\lambda_n(H_\gamma(T), C(T)) = d^n(H_\gamma(T), C(T)) = i_n(H_\gamma(T), C(T))$$

$$= \begin{cases} \left(2 \sum_{k=m}^{\infty} \gamma_k^{-1}\right)^{1/2}, & n = 2m-1, \\ \left(\gamma_m^{-1} + 2 \sum_{k=m+1}^{\infty} \gamma_k^{-1}\right)^{1/2}, & n = 2m. \end{cases}$$

The examples of $H_\gamma(T)$ are the periodic Hardy–Sobolev and Bergman–Sobolev classes ($H_2^{r,\beta}(T)$ and $A_2^{r,\beta}(T)$) which were defined in Section 1.4. For $H_2^{r,\beta}(T)$

$$\gamma_k = k^{2r} \cosh 2k\beta, \quad k = 0, \pm 1, \ldots ,$$

and for $A_2^{r,\beta}(T)$

$$\gamma_0 = 1, \qquad \gamma_k = \frac{1}{2\beta} k^{2r-1} \sinh 2k\beta, \quad k = \pm 1, \pm 2, \ldots .$$

From Theorem 4.30 we obtain

Corollary 4.31. *For all* $n \in N$

$$\lambda_n(H_2^{r,\beta}(T), C(T)) = d^n(H_2^{r,\beta}(T), C(T)) = i_n(H_2^{r,\beta}(T), C(T))$$

$$= \begin{cases} \left(\sum_{k=m}^{\infty} \frac{2}{k^{2r} \cosh 2k\beta}\right)^{1/2}, & n = 2m-1, \\ \left(\frac{1}{m^{2r} \cosh 2m\beta} + \sum_{k=m+1}^{\infty} \frac{2}{k^{2r} \cosh 2k\beta}\right)^{1/2}, & n = 2m, \end{cases}$$

$$\lambda_n(A_2^{r,\beta}(T), C(T)) = d^n(A_2^{r,\beta}(T), C(T)) = i_n(A_2^{r,\beta}(T), C(T))$$

$$= \begin{cases} \left(\sum_{k=m}^{\infty} \frac{4\beta}{k^{2r-1} \sinh 2k\beta}\right)^{1/2}, & n = 2m-1, \\ \left(\frac{2\beta}{m^{2r-1} \sinh 2m\beta} + \sum_{k=m+1}^{\infty} \frac{4\beta}{k^{2r-1} \sinh 2k\beta}\right)^{1/2}, & n = 2m. \end{cases}$$

Let us compare the values $s_n(H_2^\beta(T), C(T))$ which can be found From Theorem 2.18 with the corresponding information n-widths. In view of Corollary 4.31 we obtain the following asymptotic equalities

$$i_n^2(H_2^\beta(T), C(T)) = \begin{cases} \dfrac{4e^{-\beta}}{1-e^{-2\beta}} e^{-n\beta} + O(e^{-3n\beta}), & n = 2m-1, \\ 2\dfrac{1+e^{-2\beta}}{1-e^{-2\beta}} e^{-n\beta} + O(e^{-3n\beta}), & n = 2m. \end{cases}$$

For $s_n^2(H_2^\beta(T), C(T))$ we have

$$s_n^2(H_2^\beta(T), C(T)) = \begin{cases} \dfrac{8K}{\pi}\kappa(\beta) e^{-n\beta} + O(e^{-3n\beta}), & n = 2m-1, \\ \dfrac{8K}{\pi} e^{-n\beta} + O(e^{-3n\beta}), & n = 2m, \end{cases}$$

where K is the complete elliptic integral of the first kind for the moduli $\kappa(\beta)$. Using the asymptotic of $\kappa(\beta)$ (see (A.39)) it can be shown that

$$\lim_{n\to\infty} \frac{s_{2n-1}(H_2^\beta(T), C(T))}{i_{2n-1}(H_2^\beta(T), C(T))} = 2 - e^{-2\beta} + O(e^{-4\beta}),$$

$$\lim_{n\to\infty} \frac{s_{2n}(H_2^\beta(T), C(T))}{i_{2n}(H_2^\beta(T), C(T))} = \sqrt{2} + O(e^{-4\beta}).$$

In contrast to the class $H_{\infty,R}^\beta(T)$, where s_n and i_n are differ only for odd n (see Theorem 4.15), for the class $H_2^\beta(T)$ these quantities differ in the both cases even and odd n.

Now we consider the Bernstein n-widths. We need the following simple property of b_n.

Lemma 4.32. *Let H be a Hilbert space, A a closed, convex, centrally symmetric subset of H, and H_r an r-dimensional subspace of H such that $A \perp H_r$. Then*

$$b_{n+r}(A + H_r, H) = b_n(A, H). \tag{4.34}$$

Proof. Assume that $H_{n+1} \subset H$, $\dim H_{n+1} = n+1$, and

$$\sup\{\lambda \mid \lambda B H_{n+1} \subset A\} = \mu > 0.$$

Put $H_{n+r+1} := H_{n+1} + H_r$. Since $A \perp H_r$ it follows that $H_{n+1} \perp H_r$ and $\dim H_{n+r+1} = n + r + 1$. If $x \in H_{n+r+1}$, $\|x\|_H \le \mu$, then $x = x_1 + x_2$, where $x_1 \in H_{n+1}$, $x_2 \in H_r$, and

$$\|x_1\|_H^2 \le \|x_1\|_H^2 + \|x_2\|_H^2 = \|x\|_H^2 \le \mu^2.$$

Thus $x_1 \in A$. Consequently, $x \in A + H_r$. Hence

$$\sup\{\, \lambda \mid \lambda B H_{n+r+1} \subset A + H_r \,\} \geq \mu.$$

So we proved that

$$b_{n+r}(A + H_r, H) \geq b_n(A, H). \tag{4.35}$$

If $b_{n+r}(A + H_r, H) = 0$, then (4.34) follows from (4.35). Suppose that $b_{n+r}(A + H_r, H) > 0$. Let $H_{n+r+1} \subset H$, $\dim H_{n+r+1} = n + r + 1$, and

$$\sup\{\, \lambda \mid \lambda B H_{n+r+1} \subset A + H_r \,\} = \nu > 0.$$

Since $H_{n+r+1} \subset \operatorname{span} A + H_r$, $\dim(H_{n+r+1} \cap \operatorname{span} A) \geq n + 1$. Hence there exists a subspace $H_{n+1} \subset H_{n+r+1} \cap \operatorname{span} A$ such that $\dim H_{n+1} = n + 1$. Let $x \in H_{n+1}$ and $\|x\|_H \leq \nu$. Then $x \in A + H_r$. Moreover, $x \in \operatorname{span} A$. Consequently, $x \in A$ and

$$\sup\{\, \lambda \mid \lambda B H_{n+1} \subset A \,\} \geq \nu.$$

Therefore,

$$b_n(A, H) \geq b_{n+r}(A + H_r, H).$$

□

Put

$$D_\rho := \{\, z \in C : |z| \leq \rho \,\}.$$

Denote by ν_ρ the normalized Lebesgue measure on D_ρ.

Theorem 4.33. *For all $n \geq r$ and $0 < \rho \leq 1$*

$$b_n(H^r_\infty, L_2(T_\rho, \sigma_\rho)) = \left(\sum_{k=r}^{n} \left(\frac{k!}{(k-r)!} \right)^2 \rho^{-2k} \right)^{-1/2}, \tag{4.36}$$

$$b_n(H^r_\infty, L_2(D_\rho, \nu_\rho)) = \left(\sum_{k=r}^{n} \frac{k!(k+1)!}{((k-r)!)^2} \rho^{-2k} \right)^{-1/2}. \tag{4.37}$$

Proof. For $f(z) = \sum_{k=0}^{\infty} a_k z^k \in \mathcal{H}_2$ define the operator $T_0 \colon \mathcal{H}_2 \to L_2(T_\rho, \sigma_\rho)$ by

$$(T_0 f)(z) := \sum_{k=0}^{\infty} \frac{k!}{(k+r)!} a_k z^{k+r}.$$

Define the operator $T\colon \mathcal{H}_\infty \to L_2(T_\rho, \sigma_\rho)$ by the equation $Tf := T_0 f$. Since $(Tf)^{(r)}(z) = f(z)$ we have

$$H^r_\infty = T(B\mathcal{H}_\infty) + \mathcal{P}_r.$$

The functions $\{\varphi_k(z)\}_0^\infty := \{z^k\}_0^\infty$ form an orthogonal system in $L_2(T_\rho, \sigma_\rho)$. By Lemma 4.32 we obtain

$$b_n(H^r_\infty, L_2(T_\rho, \sigma_\rho)) = b_{n-r}(T). \tag{4.38}$$

To find $b_{n-r}(T)$ we use Theorem 4.22. The system $\{\varphi_k\}_0^\infty$ is an orthonormal basis in $\mathcal{H}_2$. Moreover, the functions

$$(T_0\varphi_k)(z) = \frac{k!}{(k+r)!} z^{k+r}$$

form an orthogonal system in $L_2(T_\rho, \sigma_\rho)$. Consequently, by Lemma 4.16

$$T_0' T_0 \varphi_k = s_k^2 \varphi_k, \quad k = 0, 1, \dots\,,$$

where

$$s_k = \|T_0\varphi_k\|_{L_2(T_\rho, \sigma_\rho)} = \frac{k!}{(k+r)!}\rho^{k+r}.$$

Using Theorem 4.22, we obtain

$$b_{n-r}(T) = \left(\sum_{k=0}^{n-r} \left(\frac{(k+r)!}{k!}\right)^2 \rho^{-2(k+r)}\right)^{-1/2}.$$

It is easily seen that this quantity coincides with the left-hand side of (4.36).

The proof of (4.37) is similar. The difference only that

$$\|T_0\varphi_k\|_{L_2(D_\rho, \nu_\rho)} = \frac{k!}{(k+r)!}\frac{\rho^{k+r}}{\sqrt{k+r+1}}.$$

□

The periodic analogue of Theorem 4.33 is the following

Theorem 4.34. *For all $n, r \in N$*

$$b_{2n}(H^{r,\beta}_\infty(T), L_2(T)) = \left(\frac{1}{\pi}\sum_{k=1}^{n} k^{2r}\cosh 2k\beta\right)^{-1/2}.$$

If $r = 0$, then for all $n \geq 0$

$$b_{2n}(H^{\beta}_\infty(T), L_2(T)) = \left(\frac{2\pi \sinh\beta}{\sinh(2n+1)\beta}\right)^{1/2}.$$

Proof. Let $r \geq 1$. Denote by $\mathcal{H}^{\beta}_{p,0}(T)$ the subspace of functions from $\mathcal{H}^{\beta}_{p}(T)$ for which

$$\int_0^{2\pi} f(t)\,dt = 0.$$

Define the operator $T_0\colon \mathcal{H}^{\beta}_{2,0}(T) \to L_2(T)$ by the equation

$$(T_0 f)(z) := \sum_{\substack{k\in Z\\ k\neq 0}} \frac{c_k}{(ik)^r} e^{ikz},$$

where

$$f(z) = \sum_{\substack{k\in Z\\ k\neq 0}} c_k e^{ikz}.$$

Put $T := T_0\big|_{\mathcal{H}^{\beta}_{\infty,0}(T)}$. We have

$$H^{r,\beta}_{\infty}(T) = T(B\mathcal{H}^{\beta}_{\infty,0}) + C.$$

By Lemma 4.32 we obtain

$$b_{2n}(H^{r,\beta}_{\infty}(T), L_2(T)) = b_{2n-1}(T).$$

It is easy to check that the functions

$$\varphi_k(z) := \frac{e^{ikz}}{\sqrt{\cosh 2k\beta}}, \quad k = \pm 1, \pm 2, \dots\ ,$$

form an orthonormal basis in $\mathcal{H}^{\beta}_{2,0}(T)$. Furthermore, the function

$$(T_0\varphi_k)(z) = \frac{e^{ikz}}{(ik)^r\sqrt{\cosh 2k\beta}}$$

form an orthogonal system in $L_2(T)$ and

$$\|T_0\varphi_k\|_2^2 = \frac{2\pi}{k^{2r}\cosh 2k\beta}.$$

Since for $|\operatorname{Im} z| = \beta$

$$|\varphi_k(z)|^2 + |\varphi_{-k}(z)|^2 = 2,$$

it follows from Lemma 4.16 and Theorem 4.22 that

$$b_{2n-1}(T) = \left(\frac{1}{\pi}\sum_{k=1}^{n} k^{2r}\cosh 2k\beta\right)^{-1/2}.$$

For $r=0$ the analogous arguments applied to the embedding operators $T_0\colon \mathcal{H}_2^\beta(T)\to L_2(T)$ and $T\colon \mathcal{H}_\infty^\beta(T)\to L_2(T)$ give

$$b_{2n}(T)=\left(\frac{1}{2\pi}+\frac{1}{\pi}\sum_{k=1}^{n}\cosh 2k\beta\right)^{-1/2}=\left(\frac{2\pi\sinh\beta}{\sinh(2n+1)\beta}\right)^{1/2}.$$

□

4.7 Notes and References

4.1. The information n-width (i_n) was introduced in Fisher, Micchelli [1984], [1985]. In these papers it was proved that for the unit balls of Hilbert spaces of nonperiodic functions with simply connected domain of holomorphy the values i_n and s_n coincide. We show that in general it is not valid for the classes $H^\beta_{\infty,R}(T)$ and $H^\beta_2(T)$ (see Sections 4.3 and 4.6). Proposition 4.1 was proved by Osipenko, Wilderotter [1997]. For the first time the Pick–Nevanlinna Theorem was applied to the problem of exact values of n-widths by Fisher, Micchelli [1980] who proved the equality (4.9). This result was generalized by Fisher, Stessin [1991] (Theorem 4.3). However, the proof of the upper bound in this paper was incorrect. The correct proof was given later in Fisher, Stessin [1994a].

For the Gel'fand n-width Corollary 4.7 was proved by Parfenov [1985]. The direct proof of this assertion which does not use Theorem 4.3 may be found in Farkov [1990]. In the case when $p=q$ there is a more general result. Denote by H_q^r the class of all functions analytic in the unit disk D for which $f^{(r)}\in H_q$. For all $1\le q\le\infty$ the equalities

$$\begin{aligned}d_n(H_q^r,L_q(T_\rho,\sigma_\rho))&=\lambda_n(H_q^r,L_q(T_\rho,\sigma_\rho))=d^n(H_q^r,L_q(T_\rho,\sigma_\rho))\\&=i_n(H_q^r,L_q(T_\rho,\sigma_\rho))=s_n(H_q^r,L_q(T_\rho,\sigma_\rho))=\begin{cases}\infty, & n<r,\\ \dfrac{(n-r)!}{n!}\rho^{n-r}, & n\ge r\end{cases}\end{aligned}$$

hold. These equalities were obtained by Babenko [1958] ($q=\infty$, the upper bound), Tikhomirov [1960] ($q=\infty$, the lower bound), and Taikov [1967] (see also Pinkus [1985, p. 250]). Exact values of n-widths for the class of functions from H_∞^r which are real on the real axis in a more general situation were considered by Fisher [1989].

4.2. Proposition 4.8 was proved by Newman [1979]. Proposition 4.9 is due to Ganelius [1976]. Theorem 4.12 was proved by Wilderotter [1995] (previous results may be found in Wilderotter [1992a] and Burchard, Höllig [1985]).

4.3. Lemma 4.13 was proved in Osipenko [1997a]. Lemma 4.14 was obtained by Osipenko, Wilderotter [1997]. Exact values of even n-widths for the class $H^{\beta}_{\infty,R}(T)$ in the space $L_q(T)$ were obtained by Osipenko [1994c]. For the class $H^{r,\beta}_{\infty,R}(T)$ even n-widths in the space $L_\infty(T)$ were determined in Osipenko [1997a]. The odd dimensional case was studied in Osipenko, Wilderotter [1997]. A general approach which generalizes Theorem 4.15 and enables one to obtain exact values of n-widths both for classes of smooth and analytic functions was proposed in Osipenko [1997b].

Denote by $\widetilde{B}^{\beta}_{2n}$ the set of the functions

$$B(z) = k^n \prod_{j=1}^{2n} \operatorname{sn}\left(\frac{K}{\pi}(z-t_j)\right),$$

where $k = \kappa(\beta)$ (the function κ is defined by (2.5)) and $t_j \in T$, $j = 1, \dots, 2n$ ($B(z)$ is the Blaschke product for the strip $S_\beta := \{z \in C : |\operatorname{Im} z| < \beta\}$ in the periodic case). Let $H^{\beta}_{p,R}$ be the class of all functions which are analytic and 2π-periodic in S_β, real-valued on the real axis, and satisfy the condition

$$\sup_{0 \le \eta < \beta} \frac{1}{2\pi} \int_0^{2\pi} |f(t+i\eta)|^p \, dt \le 1.$$

Wilderotter [1996] proved the following result which is the analogue of Theorem 4.3.

Theorem 4.35. *For all* $1 \le q \le p \le \infty$

$$\begin{aligned} d_{2n}(H^{\beta}_{p,R}, L_q(T,\mu)) = \lambda_{2n}(H^{\beta}_{p,R}, L_q(T,\mu)) &= d^{2n}(H^{\beta}_{p,R}, L_q(T,\mu)) \\ &= \inf_{B \in \widetilde{B}^{\beta}_{2n}} \sup_{g \in H^{\beta}_{p,R}} \|Bg\|_{L_q(T,\mu)}, \end{aligned}$$

where μ *is a positive measure on* T.

4.4. Theorem 4.17 was proved by Parfenov [1994b]. Theorem 4.20 is dual to the following result which is due to Ismagilov [1968]:

Theorem 4.36. *Let* E *be a compact,* φ: $E \to H$ *a continuous mapping,* H *a Hilbert space, and* μ *a probability measure on* E *with the support coincided with* E. *Denote by* $\lambda_1 \ge \lambda_2 \ge \ldots > 0$ *the eigenvalues of the problem*

$$\int_E (\varphi(x), \varphi(y)) f(y) \, d\mu(y) = \lambda f(x), \quad x \in E.$$

Let $f_1(x), f_2(x), \dots$ be associated eigenfunctions which form an orthonormal system in $L_2(E, \mu)$. Then

$$\sqrt{\sum_{k=n+1}^{\infty} \lambda_k} \leq d_n(\varphi(E), H) \leq \sup_{x \in E} \sqrt{\sum_{k=n+1}^{\infty} \lambda_k |f_k(x)|^2}.$$

Osipenko [1995b] proved Theorem 4.20 using Ismagilov's Theorem. We give the proof of Parfenov [1994b] who obtained the same result independently using extremal properties of the s-numbers. Theorem 4.22 was proved by Osipenko, Parfenov [1995].

4.5. Theorem 4.23 in the finite-dimensional case for $a_1 = \dots = a_N = 1$ was, in fact, proved in the paper of Kolmogorov, Petrov, SmirnovSmirnov, Yu. M. [1947] which was supplemented by Mal'cev [1947]. The authors of these papers did not actually state, that they had calculated n-widths. This was noted by Stechkin [1954]. The case $a_1 \geq \dots \geq a_N$ was obtained by Smolyak [1965]. The dual result related to the Kolmogorov n-width of the set

$$Bl_1(a) := \left\{ (x_1, x_2, \dots) \,\middle|\, \sum_{k=1}^{\infty} \frac{|x_k|}{a_k} \leq 1 \right\}$$

in the space l_2 is due to Sofman [1969], [1973] (see also Hutton, Morrel, Retherford [1976]). Lemmas 4.24 and 4.25 are taken from Pinkus [1985]. Theorem 4.27 (in a different way) was proved by Galeev [1990].

4.6. As was mentioned in Section 4.4, the first exact result for the classes of holomorphic functions concerned the case when $p < q$ appeared in Osipenko, Stessin [1990], where the exact values of $d^N(H_2(B_n), C(\rho S_n))$ and $\lambda^N(H_2(B_n), C(\rho S_n))$ were obtained for

$$N = N_m := \sum_{s=0}^{m-1} \binom{n+s-1}{n-1}$$

(the definition of $H_2(B_n)$ is given in Section 2.5). The method of proof, as noted by Tikhomirov, was very similar to the one used in Ismagilov's Theorem. This remark was taken into account in Osipenko [1995b] where more general results were obtained. The classes H_γ were proposed by Fisher, Micchelli [1985] as a convenient form for the generalization of certain classes of analytic functions. The assertion of Corollary 4.29 is contained in Osipenko [1995b]. In the last paper the case of several variables was treated (in the one dimensional case the exact values of n-widths of the Bergman class A_2

were obtained by Fisher, Stessin [1994b]). Theorem 4.30 and Corollary 4.31 were proved by Osipenko [1995b].

The exact values of the Bernstein n-widths of the class H^r_∞ were obtained by Parfenov [1994a] ($r = 0$) and Osipenko, Parfenov [1995].

Chapter 5

Quadrature Formulas

There are two basic problems which are usually considered for quadrature formulas. The first one is optimal quadratures which use function values at a fixed system of points. The second one is the minimization of the error of optimal quadrature formulas by choosing appropriate system of nodes at which function values should be known. For the Hardy classes H_p the first problem is considered in Section 5.1. In Section 5.2 we find optimal nodes for the periodic case and for some special weight functions in the nonperiodic case. In particular, we prove that in the periodic case the equidistant system of nodes is the only optimal to within a translation. In Section 5.3 we obtain estimates of optimal quadrature formulas in H_p for all $1 < p \leq \infty$ in the case when the integral is taken over the interval $(-1, 1)$.

5.1 Best quadrature formulas in H_p

Let W be a class of functions defined on the domain $G \in C$ containing an interval of real axis (a, b). For this class we consider the problem of optimal recovery of the functional

$$Lf := \int_a^b f(x)p(x)\,dx$$

from the values of the information operator

$$If := \{\, f(x_1), \dots, f^{(\nu_1-1)}(x_1), \dots, f(x_n), \dots, f^{(\nu_n-1)}(x_n) \,\},$$

where p is a nonnegative weight function and $x_1, \dots, x_n$ are distinct nodes of the set $R \cap G$ (we assume that functions from W are sufficiently smooth).

For $\nu := (\nu_1, \dots, \nu_n)$ set

$$\tau_\nu := \begin{pmatrix} x_1 & \dots & x_n \\ \nu_1 & \dots & \nu_n \end{pmatrix}, \tag{5.1}$$

$$e(\tau_\nu, W) := \inf_{\varphi\colon C^n \to C} \sup_{f \in W} \left| \int_a^b f(x)p(x)\,dx - \varphi(If) \right|.$$

If W is a convex balanced class of function, then by Theorem 1.6 we have

$$e(\tau_\nu, W) = \sup_{\substack{f \in W \\ If=0}} \left| \int_a^b f(x)p(x)\,dx \right|. \tag{5.2}$$

Moreover, there exists a linear optimal method. In other words, there exists a quadrature formula

$$\int_a^b f(x)p(x)\,dx \approx \varphi_0(If) = \sum_{j=1}^{n} \sum_{m=0}^{\nu_j - 1} a_{jm} f^{(m)}(x_j),$$

for which

$$\sup_{f \in W} \left| \int_a^b f(x)p(x)\,dx - \varphi_0(If) \right| = e(\tau_\nu, W).$$

Such quadrature formulas are said to be *best for a given system of nodes* τ_ν.

Let $(a, b) \subset (-1, 1)$, $x_j \in (-1, 1)$, $j = 1, \dots, n$. Set

$$B(x) := \prod_{j=1}^{n} \left(\frac{x - x_j}{1 - x_j x} \right)^{\nu_j}.$$

Lemma 5.1. *Assume that* $\nu_1, \dots, \nu_n$ *are even,* $1 < p < \infty$ *or* $p = 1$ *and* $-1 < a < b < 1$. *Then*

(i) there exists a unique function $g_{\tau_\nu, p} \in H_p$ *such that*

$$e(\tau_\nu, H_p) = \int_a^b g_{\tau_\nu, p}(x) B(x) p(x)\,dx, \tag{5.3}$$

(ii) $g_{\tau_\nu, p}$ *is free of zeros in the unit disk* D *and* $g_{\tau_\nu, p}(x) > 0$ *for* $x \in (-1, 1)$,

(iii) for all $\theta \in T$

$$e(\tau_\nu, H_p) |g_{\tau_\nu, p}(e^{i\theta})|^p = \int_a^b g_{\tau_\nu, p}(x) B(x) P(e^{i\theta}, x) p(x)\,dx, \tag{5.4}$$

where $P(e^{i\theta}, x)$ *is the Poisson kernel.*

Proof. It follows from Proposition 4.5 that there exists a unique function $g_B \in H_p$ such that

$$P_1 := \sup_{g\in H_p} \int_a^b |g(x)|B(x)p(x)\,dx = \int_a^b |g_B(x)|B(x)p(x)\,dx$$

and $g_B(0) > 0$. Since $\overline{g_B(\overline{z})}$ is also extremal for the problem of finding P_1, we obtain that $\overline{g_B(\overline{z})} = g_B(z)$. In particular, $g_B(x)$ is real for $x \in (-1,1)$. It was also shown in Section 4.1 that $g_B(z)$ is free of zeros in D. Thus $g_B(x) > 0$ for $x \in (-1,1)$ and

$$P_1 = P_2 := \sup_{g\in H_p} \left| \int_a^b g(x)B(x)p(x)\,dx \right|.$$

Assume that $g_0 \in H_p$ is an extremal function for the problem of finding P_2. Then

$$P_2 = \left| \int_a^b g_0(x)B(x)p(x)\,dx \right| \le \int_a^b |g_0(x)|B(x)p(x)\,dx \le P_1 = P_2.$$

Consequently, g_0 is an extremal function for the problem of finding P_1. Thus we proved (i) and (ii). Now (iii) immediately follows from Proposition 4.4. □

For $p = \infty$ and even $\nu_1, \dots, \nu_n$ it is evident that $g_{T_\nu,\infty}(z) \equiv 1$ is the unique function for which (5.3) holds.

For $p = 2$ we can also find the extremal function $g_{T_\nu,2}$. Indeed, we have

$$\begin{aligned}
&\sup_{g\in H_2} \left| \int_a^b g(x)B(x)p(x)\,dx \right| \\
&\quad = \sup_{g\in H_2} \left| \int_a^b \frac{1}{2\pi} \int_0^{2\pi} \frac{g(e^{i\theta})}{1 - xe^{-i\theta}}\,d\theta B(x)p(x)\,dx \right| = \sup_{g\in H_2} |(g,\varphi)_{\mathcal{H}_2}|,
\end{aligned}$$

where

$$\varphi(z) = \int_a^b \frac{B(x)}{1 - xz} p(x)\,dx$$

and

$$(g,\varphi)_{\mathcal{H}_2} = \frac{1}{2\pi} \int_0^{2\pi} g(e^{i\theta})\overline{\varphi(e^{i\theta})}\,d\theta$$

is the inner product in the Hilbert space $\mathcal{H}_2$. Thus

$$g_{T_\nu,2}(z) = \frac{\varphi(z)}{\|\varphi\|_{\mathcal{H}_2}}.$$

Theorem 5.2. *Let $(a,b) \subset (-1,1)$, $x_j \in (-1,1)$, $j = 1,\dots,n$. Assume that $1 < p \le \infty$ or $p = 1$ and $-1 < a < b < 1$. Then for even ν_j, $j = 1,\dots,n$, the quadrature formula*

$$\int_a^b f(x)p(x)\,dx \approx \sum_{j=1}^{n}\sum_{m=0}^{\nu_j-1} a_{jm} f^{(m)}(x_j),$$

where

$$a_{jm} = \int_a^b c_{jm}(x)p(x)\,dx,$$

$$c_{jm}(x) = \frac{B(x)g_{\tau_\nu,p}(x)(1-x^2)}{m!(\nu_j-m-1)!}\left(\frac{(1-x_jz)^{\nu_j}}{\omega_j(z)g_{\tau_\nu,p}(z)(x-z)(1-xz)}\right)^{(\nu_j-m-1)}_{\Big|z=x_j},$$

$$\omega_j(x) = \prod_{\substack{s=1\\ s\ne j}}^{n}\left(\frac{x-x_s}{1-x_sx}\right)^{\nu_s},$$

is best on the class H_p for the system of nodes τ_ν.

Proof. For $x \in (a,b)$ consider the integral

$$Jf(x) := B(x)g_{\tau_\nu,p}(x)\frac{1}{2\pi i}\int_{|z|=1}\frac{f(z)(1-x^2)}{B(z)g_{\tau_\nu,p}(z)(z-x)(1-xz)}\,dz.$$

By the residue theorem we get

$$Jf(x) = f(x) - \sum_{j=1}^{n}\sum_{m=0}^{\nu_j-1} c_{jm}(x)f^{(m)}(x_j).$$

It is easy to verify that

$$Jf(x) = B(x)g_{\tau_\nu,p}(x)\int_0^{2\pi}\frac{f(e^{i\theta})}{B(e^{i\theta})g_{\tau_\nu,p}(e^{i\theta})}\,d\nu(\theta),$$

where

$$d\nu(\theta) = \frac{1}{2\pi}P(e^{i\theta},x)\,d\theta.$$

Assume that $p < \infty$. Using (5.4) we obtain

$$R_f := \left|\int_a^b f(x)p(x)\,dx - \sum_{j=1}^{n}\sum_{m=0}^{\nu_j-1} a_{jm}f^{(m)}(x_j)\right| = \left|\int_a^b Jf(x)p(x)\,dx\right|$$
$$\leq \int_a^b |Jf(x)|p(x)\,dx \leq \int_a^b B(x)g_{\tau_\nu,p}(x)\int_0^{2\pi}\frac{|f(e^{i\theta})|}{|g_{\tau_\nu,p}(e^{i\theta})|}\,d\nu(\theta)p(x)\,dx$$
$$= \frac{1}{2\pi}\int_0^{2\pi}\frac{|f(e^{i\theta})|}{|g_{\tau_\nu,p}(e^{i\theta})|}\int_a^b g_{\tau_\nu,p}(x)B(x)P(e^{i\theta},x)p(x)\,dx\,d\theta$$
$$= e(\tau_\nu, H_p)\frac{1}{2\pi}\int_0^{2\pi}|f(e^{i\theta})||g_{\tau_\nu,p}(e^{i\theta})|^{p-1}\,d\theta.$$

By the Hölder inequality we have

$$R_f \leq e(\tau_\nu, H_p)\left(\int_0^{2\pi}|f(e^{i\theta})|^p\,d\theta\right)^{1/p}\left(\int_0^{2\pi}|g_{\tau_\nu,p}(e^{i\theta})|^p\,d\theta\right)^{(p-1)/p}$$
$$\leq e(\tau_\nu, H_p).$$

If $p = \infty$, then $g_{\tau_\nu,\infty} \equiv 1$ and

$$R_f \leq \int_a^b B(x)p(x)\,dx = e(\tau_\nu, H_\infty).$$

□

Now we consider the periodic case. For

$$\tau_\nu := \begin{pmatrix} t_1 & \dots & t_n \\ \nu_1 & \dots & \nu_n \end{pmatrix},$$

where $t_j \in T$, $j = 1, \dots, n$, set

$$N := \sum_{j=1}^{n}\nu_j, \qquad B(z) := k^{\frac{N}{2}}\prod_{j=1}^{n}\operatorname{sn}^{\nu_j}\left(\frac{K}{\pi}(z - t_j)\right)$$

(K is the complete elliptic integral of the first kind for the modulus $k = \kappa(\beta)$ and κ is defined by (2.5)).

Theorem 5.3. *Let* $(a, b) \subset (0, 2\pi)$ *and* ν_j *be even. Then the quadrature formula*

$$\int_a^b f(t)p(t)\,dt \approx \sum_{j=1}^{n}\sum_{m=0}^{\nu_j-1} a_{jm}f^{(m)}(t_j), \tag{5.5}$$

in which

$$a_{jm} = \int_a^b c_{jm}(t)p(t)\,dt,$$

$$c_{jm}(t) = \frac{K}{\pi}\frac{B(t)}{m!(\nu_j - m - 1)!}\lim_{z\to t_j}\left(\frac{(z-t_j)^{\nu_j}}{B(z)}\operatorname{ctn}\left(\frac{K}{\pi}(t-z)\right)\right)^{(\nu_j-m-1)},$$

$$\operatorname{ctn} z = \frac{\operatorname{cn} z \operatorname{dn} z}{\operatorname{sn} z},$$

is best on the class $H^\beta_\infty(T)$ *for the system of nodes* τ_ν. *Moreover,*

$$e(\tau_\nu, H^\beta_\infty(T)) = \int_a^b B(t)p(t)\,dt. \tag{5.6}$$

Proof. It follows from Theorem 2.3 that for all $t \in T$ the method

$$f(t) \approx \sum_{j=1}^{n}\sum_{m=0}^{\nu_j-1} c_{jm}(t)f^{(m)}(t_j)$$

is optimal on the class $H^\beta_\infty(T)$ and its intrinsic error is equal to $B(t)$ (since ν_j are even, $B(t) \geq 0$). Thus for all $t \in (a,b)$ and all $f \in H^\beta_\infty(T)$

$$\left|f(t) - \sum_{j=1}^{n}\sum_{m=0}^{\nu_j-1} c_{jm}(t)f^{(m)}(t_j)\right| \leq B(t).$$

Hence

$$\left|\int_a^b f(t)p(t)\,dx - \sum_{j=1}^{n}\sum_{m=0}^{\nu_j-1} a_{jm}f^{(m)}(t_j)\right| \leq \int_a^b B(t)p(t)\,dt. \tag{5.7}$$

Consequently,

$$e(\tau_\nu, H^\beta_\infty(T)) \leq \int_a^b B(t)p(t)\,dt.$$

On the other hand, since $B \in H^\beta_\infty(T)$ and $B^{(m)}(t_j) = 0$, $j = 1,\dots,n$, $m = 0,\dots,\nu_j - 1$, it follows from (5.2) that

$$e(\tau_\nu, H^\beta_\infty(T)) \geq \int_a^b B(t)p(t)\,dt.$$

Thus (5.6) is proved. Now it follows from (5.7) that the quadrature formula (5.5) is best. □

5.2 Optimal Quadrature Formulas

Now we would like to find a system of nodes at which the error of the best quadrature formula is minimal. Put

$$e(\nu, W) := \inf_{\substack{x_1<\ldots<x_n \\ x_j \in G\cap R}} e(\tau_\nu, W), \tag{5.8}$$

where G is a domain of definition of functions from the class W. The nodes τ_ν at which the lower bound in (5.8) is attained are called *optimal nodes of type* ν and the best quadrature formula for the system of optimal nodes is said to be *optimal.*

Theorem 5.4. *Let* $(a,b) = (0,2\pi)$, $p(t) \equiv 1$, *and* $s \in N$. *Then for all* $2s-1 \le \nu_j \le 2s$, $j = 1,\ldots,n$,

$$e(\nu, H_\infty^\beta(T)) = \frac{2\pi\lambda^s}{\Lambda} J_{2s}(\lambda) = 2^{s+1}\pi\frac{(2s-1)!!}{s!}e^{-s\beta n} + O\left(e^{-(s+2)\beta n}\right),$$

where Λ *is the complete elliptic integral of the first kind for the modulus* $\lambda = \kappa(\beta n)$ *and*

$$J_q(\lambda) := \int_0^1 \frac{t^q\,dt}{\sqrt{(1-t^2)(1-\lambda^2 t^2)}}. \tag{5.9}$$

Moreover, the equidistant system of nodes on T *is the only system of optimal nodes to within a translation.*

Proof. Let $\mu = (2s\ldots 2s)$. It follows from Theorem 5.3 that

$$\begin{aligned} e(\mu, H_\infty^\beta(T)) &= \inf_{0\le t_1<\ldots<t_n<2\pi}\int_0^{2\pi} B(t)\,dt \\ &= \inf_{0\le t_1<\ldots<t_n<2\pi}\left\|k^{n/2}\prod_{j=1}^n \operatorname{sn}\left(\frac{K}{\pi}(\cdot - t_j)\right)\right\|_{2s}^{2s}. \end{aligned}$$

Putting in Theorem 2.14 $\varphi(x) = \tan\frac{\pi}{4}x$, we obtain

$$e(\mu, H_\infty^\beta(T)) = \frac{2\pi\lambda^s}{\Lambda}J_{2s}(\lambda). \tag{5.10}$$

It follows from the same theorem that the equidistant system of nodes on T is the only system of optimal nodes to within a translation. The asymptotic equality for the value (5.10) follows easily from (2.34).

For optimal nodes $t_j^* = (j-1)\dfrac{2\pi}{n}$, $j = 1, \dots, n$, we consider the quadrature formula (5.5). We denote by B_0 the function B for this system of nodes. We have

$$a_{j,2s-1} = \left(\frac{\pi}{K}\right)^{2s-1} \frac{1}{(2s-1)!k^s \omega_{j0}(t_j^*)} \int_0^{2\pi} B_0(t) \operatorname{ctn}\left(\frac{K}{\pi}(t - t_j^*)\right) dt,$$

where

$$\omega_{j0}(t) = k^{s(n-1)} \prod_{\substack{m=1\\ m\neq j}}^{n} \operatorname{sn}^{2s}\left(\frac{K}{\pi}(t - t_m^*)\right).$$

On the other hand, it follows from the necessary condition of extremum that

$$\int_0^{2\pi} B_0(t) \operatorname{ctn}\left(\frac{K}{\pi}(t - t_j^*)\right) dt = 0.$$

Thus $a_{j,2s-1} = 0$ for all $j = 1, \dots, n$. Hence for all $(\nu_1, \dots, \nu_n)$ such that $2s - 1 \leq \nu_j \leq 2s$, $j = 1, \dots, n$,

$$e(\nu, H_\infty^\beta(T)) \leq \int_0^{2\pi} B_0(t)\, dt.$$

If $t_1 \dots, t_n$ is an arbitrary system of nodes with multiplicities $2s - 1 \leq \nu_j \leq 2s$, $j = 1, \dots, n$, then from (5.2) we have

$$e(\tau_\nu, H_\infty^\beta(T)) \geq \left\| k^{n/2} \prod_{j=1}^{n} \operatorname{sn}\left(\frac{K}{\pi}(\cdot - t_j)\right) \right\|_{2s}^{2s} \geq \int_0^{2\pi} B_0(t)\, dt.$$

In view of uniqueness the last inequality is strong provided that $t_1 \dots, t_n$ is not the equidistant system on T. □

Now consider the problem of optimal quadrature formula for the integral

$$\int_{-1}^{1} f(t) \frac{dt}{\sqrt{1-t^2}} \tag{5.11}$$

on the class $H_\infty(\mathcal{E}_c)$ which is the set of all functions analytic in the interior of the ellipse $\mathcal{E}_c$ with foci at the points ± 1 and sum $c > 1$ of its semiaxes and satisfying the condition $|f(z)| \leq 1$, $z \in \mathcal{E}_c$.

Theorem 5.5. *For all $c > 1$, $s \in N$, and $2s - 1 \leq \nu_j \leq 2s$, $j = 1, \dots, n$,*

$$e(\nu, H_\infty(\mathcal{E}_c)) = \frac{\pi\lambda^s}{\Lambda} J_{2s}(\lambda) = 2^s \pi \frac{(2s-1)!!}{s!} c^{-2sn} + O\left(c^{-2(s+2)n}\right),$$

where Λ is the complete elliptic integral of the first kind for the modulus $\lambda = \kappa(2n \log c)$. Moreover, the Chebyshev nodes

$$x_j^* = \cos\frac{2j-1}{2n}\pi, \quad j = 1, \dots, n, \tag{5.12}$$

are the only optimal nodes.

Proof. By the conformal mapping of the interior of the ellipse $\mathcal{E}_c$ on the unit disk defined by (2.35) the original problem reduces to the one of finding the error of optimal quadrature formula on the class H_∞ for the integral

$$\frac{\pi}{2K} \int_{-\sqrt{k}}^{\sqrt{k}} f(x) \frac{dx}{\sqrt{(k - x^2)(1 - kx^2)}},$$

where K is the complete elliptic integral of the first kind for the modulus $k = \kappa(2 \log c)$. Further, we use the same scheme as in the previous theorem but instead of referring to Theorems 5.3 and 2.14 we refer to Theorems 5.2 and 2.15. □

We take up more explicitly the case when $\nu_1 = \dots = \nu_n = 1$. Assume that for any system of nodes $\tau = (x_1, \dots, x_n)$ the information operator

$$I_\tau f := (f(x_1), \dots, f(x_n))$$

is known with an error at most δ in the norm of the space l_p^n, $1 \leq p \leq \infty$, that is, we know the values $\widetilde{f} = (\widetilde{f}(x_1), \dots, \widetilde{f}(x_n))$ which satisfy the condition

$$\|I_\tau f - \widetilde{f}\|_{l_p^n} := \left(\sum_{j=1}^n |f(x_j) - \widetilde{f}(x_j)|^p\right)^{1/p} \leq \delta.$$

We shall be interested in the problem of finding the value

$$e_{np}(W, \delta) := \inf_{x_j \in G \cap R} \inf_{a_j} \sup_{f \in W} \sup_{\substack{\widetilde{f} \in l_p^n \\ \|I_\tau f - \widetilde{f}\|_{l_p^n} \leq \delta}} \left| \int_a^b f(x) p(x)\, dx - \sum_{j=1}^n a_j \widetilde{f}(x_j) \right|. \tag{5.13}$$

A quadrature formula for which the lower bound in (5.13) is attained and corresponding nodes are called optimal as before.

For a convex balanced class W by Theorem 1.6 we have

$$e_{np}(W,\delta) = \inf_{x_j \in G\cap R} \sup_{\substack{f\in W \\ \|I_\tau f\|_{l_p^n}\le\delta}} \left| \int_a^b f(x)p(x)\,dx \right|. \tag{5.14}$$

If $W = H_\infty^\beta(T)$ and $\delta \ge n^{1/p}$ it is easy to see that the function $f(x) \equiv 1$ is extremal in (5.14) for any system of nodes and, consequently, the quadrature formula with $a_j = 0$, $j = 1, \dots, n$ is optimal for this trivial case.

Put

$$J_q(\lambda,\Delta) := \int_0^1 \left(\frac{\lambda t^2 + \Delta}{1 + \Delta\lambda t^2} \right)^{q/2} \frac{dt}{\sqrt{(1-t^2)(1-\lambda^2 t^2)}}.$$

For $\Delta = 0$ this function coincides with $J_q(\lambda)$ defined by (5.9).

Theorem 5.6. *Let $1 \le p \le \infty$ and $0 \le \delta < n^{1/p}$. Then*

1) the quadrature formula

$$\int_0^{2\pi} f(t)\,dt \approx \frac{2\pi}{n} \frac{\Lambda - J_4(\lambda,\Delta)}{(1-\Delta^2)\Lambda} \sum_{j=0}^{n-1} \tilde{f}\left(j\frac{2\pi}{n} \right), \tag{5.15}$$

in which $\Delta = \delta n^{-1/p}$, is optimal on the class $H_\infty^\beta(T)$; here Λ is the complete elliptic integral of the first kind for the modulus $\lambda = \kappa(\beta n)$;

2) the following equalities hold:

$$\begin{aligned} e_{np}(H_\infty^\beta(T),\delta) &= \frac{2\pi}{\Lambda} J_2(\lambda,\Delta) \\ &= 2\pi\Delta + 4\pi(1-\Delta^2)e^{-\beta n} + 4\pi\Delta(4-3\Delta^2)e^{-2\beta n} + O\left(e^{-3\beta n}\right); \end{aligned}$$

3) the nodes $t_j^* = (j-1)\dfrac{2\pi}{n}$, $j = 1, \dots, n$, are the only optimal to within a translation.

Proof. Let $0 \le t_1 < \ldots < t_n < 2\pi$. For $k = \kappa(\beta)$ set

$$B(z) := k^{n/2} \prod_{j=1}^{n} \operatorname{sn}\left(\frac{K}{\pi}(z - t_j) \right).$$

It follows from the proof of Theorem 2.3 that $B^2(z) \in H^\beta_\infty(T)$. Moreover, $|B(x \pm i\beta)| = 1$ for all $x \in R$. For $t \in T$ and $f \in H^\beta_\infty(T)$ consider the integral

$$Jf := \frac{1}{2\pi i}\int_{\Gamma_\varepsilon} \frac{K}{\pi}\,\frac{B^2(t)(1+\Delta B^2(\xi))^2}{B^2(\xi)(1+\Delta B^2(t))^2}\operatorname{ctn}\left(\frac{K}{\pi}(\xi - t)\right) f(\xi)\,d\xi,$$

where Γ_ε is the boundary of the rectangle: $-\varepsilon \le \operatorname{Re} z \le 2\pi - \varepsilon$, $|\operatorname{Im} z| \le \beta$; here $\varepsilon > 0$ is chosen from the requirement that the points $t, t_1, \dots, t_n$ lie inside the rectangle. By the residue theorem we have

$$Jf = f(t) - \sum_{j=1}^{n}(c_{j0}(t)f(t_j) + c_{j1}(t)f'(t_j)), \tag{5.16}$$

where

$$c_{j1}(t) = \frac{K}{\pi}\,\frac{B^2(t)}{B'^2(t_j)(1+\Delta B^2(t))^2}\operatorname{ctn}\left(\frac{K}{\pi}(t - t_j)\right),$$

$$c_{j0}(t) = \frac{K^2}{\pi^2}\,\frac{B^2(t)}{B'^2(t_j)(1+\Delta B^2(t))^2}\,\frac{1 - k^2\operatorname{sn}^4\left(\frac{K}{\pi}(t-t_j)\right)}{\operatorname{sn}^2\left(\frac{K}{\pi}(t-t_j)\right)} - \frac{B''(t_j)}{B'(t_j)}c_{j1}(t).$$

For a fixed $t \in T$ consider the function

$$F(\xi) := \frac{K}{\pi}\,\frac{B^2(t)}{B^2(\xi)(1+\Delta B^2(t))^2}\operatorname{ctn}\left(\frac{K}{\pi}(\xi - t)\right).$$

The function F is an elliptic function with periods 2π and $i2\pi K'/K$. To within the periods, all the poles of F are at the points $t, t_1, \dots, t_n$, and $t + i\pi K'/K$. Using the fact that the sum of the residues of an elliptic function with respect to all the poles lying inside a period parallelogram is equal to zero, we obtain

$$\frac{1}{(1+\Delta B^2(t))^2} - \sum_{j=1}^{n} c_{j0}(t) - \frac{B^4(t)}{(1+\Delta B^2(t))^2} = 0. \tag{5.17}$$

By virtue of the 2π-periodicity of the functions $B^2(z)$ and $\operatorname{ctn}\frac{K}{\pi}z$, we can replace Γ_ε by $\Gamma := [2\pi + i\beta, i\beta] \cup [-i\beta, 2\pi - i\beta]$ in the integral Jf. It follows from (2.6) that

$$\operatorname{ctn}\left(\frac{K}{\pi}(x \pm i\beta)\right) = \mp i\left|\operatorname{dn}\left(\frac{K}{\pi}(x \pm i\beta)\right)\right|^2.$$

Thus, the integral Jf can be written in the form

$$Jf = \int_{\Gamma_0} \overline{g(\xi)}\chi(\xi,t)f(\xi)\,d\mu(\xi), \tag{5.18}$$

where

$$g(\xi) = \frac{B^2(\xi)+\Delta}{1+\Delta B^2(\xi)},$$

$$\chi(\xi,t) = \frac{KB^2(t)(1+\Delta B^2(\xi))(B^2(\xi)+\Delta)}{2\pi^2 B^2(\xi)(1+\Delta B^2(t))^2}\left|\operatorname{dn}\left(\frac{K}{\pi}(\xi-t)\right)\right|^2,$$

and $\Gamma_0 = [i\beta, 2\pi+i\beta]\cup[-i\beta, 2\pi-i\beta]$ with measure $d\mu(x\pm i\beta) = dx$. Since $|B(\xi)| = 1$ when $\xi\in\Gamma_0$, it follows that $|g(\xi)| = 1$ and $\chi(\xi,t) > 0$ for all $\xi\in\Gamma_0$ and $t\in T$. Using (5.16) and (5.18), we now obtain the representation

$$\int_0^{2\pi} f(t)\,dt - \sum_{j=1}^{n}(a_{j0}f(t_j) + a_{j1}f'(t_j)) = \int_{\Gamma_0}\overline{g(\xi)}\psi(\xi)f(\xi)\,d\mu(\xi),$$

in which

$$\psi(\xi) = \int_0^{2\pi}\chi(\xi,t)\,dt, \qquad a_{jm} = \int_0^{2\pi} c_{jm}(t)\,dt, \quad m = 0,1.$$

Denote by B_0 the function B for the nodes t_j^*. (In this case we denote the corresponding function g by g_0.) Using the first principal transform of elliptic functions of degree n (see Appendix), we can show that

$$B_0(t) = (-1)^{n+1}\sqrt{\lambda}\operatorname{sn}\left(\frac{n\Lambda}{\pi}t, \lambda\right). \tag{5.19}$$

Consequently, $B_0^2(t+t_j^*) = B_0^2(t)$ and since $\operatorname{ctn} t$ is odd, we have

$$\int_0^{2\pi}\frac{B_0^2(t)}{(1+\Delta B_0^2(t))^2}\operatorname{ctn}\left(\frac{K}{\pi}(t-t_j^*)\right)dt$$
$$= \int_{-\pi}^{\pi}\frac{B_0^2(t)}{(1+\Delta B_0^2(t))^2}\operatorname{ctn}\frac{K}{\pi}t\,dt = 0.$$

Thus, $a_{j1} = 0$, $j = 1,\dots,n$. By similar considerations we find

$$a_{j0} = \frac{K^2}{\pi^2 B_0'^2(t_j^*)}\int_0^{2\pi}\frac{B_0^2(t)}{(1+\Delta B_0^2(t))^2}\,\frac{1-k^2\operatorname{sn}^4\frac{K}{\pi}t}{\operatorname{sn}^2\frac{K}{\pi}t}\,dt.$$

From (5.19) we can obtain $B_0'(t_j^*)$ and verify that $B_0'^2(t_1^*) = \ldots = B_0'^2(t_n^*)$. Therefore $a_{10} = \ldots = a_{n0} =: A$.

To find the quantity A, we use (5.17) and integrate it to obtain

$$nA = \sum_{j=1}^{n} a_{j0} = \int_0^{2\pi} \frac{1 - B_0^4(t)}{(1 + \Delta B_0^2(t))^2}\, dt = \frac{1}{1 - \Delta^2} \int_0^{2\pi} (1 - g_0^2(t))\, dt.$$

Substituting $\frac{n\Lambda}{\pi} t = z$, $x = \operatorname{sn}(z, \lambda)$ in the latter integral, we obtain

$$A = \frac{2\pi}{n} \frac{\Lambda - J_4(\lambda, \Delta)}{(1 - \Delta^2)\Lambda}.$$

Thus, we have proved the equality

$$\int_0^{2\pi} f(t)\, dt - A \sum_{j=1}^{n} f(t_j^*) = \int_{\Gamma_0} \overline{g(\xi)} \psi(\xi) f(\xi)\, d\mu(\xi).$$

For $z = (z_1, \ldots, z_n) \in l_p^n$ we set $\langle y_0^*, z\rangle := A \sum_{j=1}^{n} z_j$. Then

$$\langle y_0^*, I_{\tau^*} g_0 \rangle = A \sum_{j=1}^{n} g_0(t_j^*) = A\delta n^{1-1/p} = \delta \|y_0^*\|$$

for $\tau^* := (t_1^*, \ldots, t_n^*)$. Furthermore, $\|I_{\tau^*} g_0\|_{l_p^n} = \delta$. Applying Theorem 1.10 we see that the quadrature formula (5.15) is the best in the class $H_\infty^\beta(T)$ for the system of nodes τ^* and the function g_0 is extremal. Consequently,

$$e_{np}(H_\infty^\beta(T), \delta) \leq \int_0^{2\pi} g_0(t)\, dt.$$

On the other hand, since for any system of nodes τ the function $g \in H_\infty^\beta(T)$ and $\|I_\tau g\|_{l_p^n} = \delta$, it follows from (5.14) that

$$e_{np}(H_\infty^\beta(T), \delta) \geq \inf_{t_j \in T} \int_0^{2\pi} g(t)\, dt = \inf_{t_j \in T} \int_0^{2\pi} \varphi\left(\frac{4}{\pi} \arctan B(t)\right) dt,$$

where

$$\varphi(x) = \frac{\tan^2 \frac{\pi}{4} x + \Delta}{1 + \Delta \tan^2 \frac{\pi}{4} x}. \tag{5.20}$$

From Theorem 2.14 we get

$$\inf_{t_j \in T} \int_0^{2\pi} g(t)\,dt = \int_0^{2\pi} \varphi\left(\frac{4}{\pi} \arctan B_0(t)\right) dt = \int_0^{2\pi} g_0(t)\,dt. \qquad (5.21)$$

Thus,

$$e_{np}(H_\infty^\beta(T), \delta) = \int_0^{2\pi} g_0(t)\,dt.$$

An expression for this quality in terms of the function J_2 is obtained with the same substitutions which were used to calculate A, while the asymptotic equality is obtained using (2.34).

Assertion 3) follows from the fact that the infimum in (5.21) is atteined only for a system of points uniformly distributed with step size $2\pi/n$. □

Now we consider the problem (5.13) for the integral (5.11) and the class $H_\infty(\mathcal{E}_c)$. For $\delta \geq n^{1/p}$ the solution of this problem is trivial and can be obtained from the same considerations used above for the class $H_\infty^\beta(T)$.

Theorem 5.7. *Let $1 \leq p \leq \infty$ and $0 \leq \delta < n^{1/p}$. Then*

1) *the quadrature formula*

$$\int_{-1}^{1} f(t) \frac{dt}{\sqrt{1-t^2}} \approx \frac{\pi}{n} \frac{\Lambda - J_4(\lambda, \Delta)}{(1-\Delta^2)\Lambda} \sum_{j=1}^{n} \tilde{f}\left(\cos \frac{2j-1}{2n}\pi\right), \qquad (5.22)$$

in which $\Delta = \delta n^{-1/p}$ and Λ is the complete elliptic integral of the first kind for the modulus $\lambda = \kappa(2n \log c)$, is optimal for the class $H_\infty(\mathcal{E}_c)$;

2) *the following equalities hold:*

$$\begin{aligned} e_{np}(H_\infty(\mathcal{E}_c), \delta) &= \frac{\pi}{\Lambda} J_2(\lambda, \Delta) \\ &= \pi\Delta + 2\pi(1-\Delta^2)c^{-2n} + 2\pi\Delta(4 - 3\Delta^2)c^{-4n} + O\left(c^{-6n}\right); \end{aligned}$$

3) *the Chebyshev nodes* (5.12) *are the only optimal nodes.*

Proof. If $f(t) \in H_\infty(\mathcal{E}_c)$, then $f(\cos t) \in H_\infty^\beta(T)$ for $\beta = \log c$. Applying Theorem 5.6, we see that the error of the quadrature formula (5.22) and, consequently, the optimal quadrature formula do not exceed the quantity

$$\frac{1}{2} e_{2n,p}(H_\infty^\beta(T), 2^{1/p}\delta) = \frac{\pi}{\Lambda} J_2(\lambda, \Delta).$$

To obtain the lower bound, we use (5.14). Using the substitution (2.35), we obtain

$$e_{np}(H_\infty(\mathcal{E}_c),\delta) = \inf_{z_j\in(-1,1)} \sup_{\substack{f\in H_\infty \\ \|I_\tau f\|_{l_p^n}\le\delta}} \left|\int_{-\sqrt{k}}^{\sqrt{k}} f(z)q(z)\,dz\right|,$$

where $\tau = (z_1,\dots,z_n)$ and

$$q(z) = \frac{\pi}{2K}\frac{1}{\sqrt{(k-z^2)(1-kz^2)}}.$$

Set

$$B(z) := \prod_{j=1}^{n}\frac{z-z_j}{1-zz_j}, \qquad g(z) := \frac{B^2(z)+\Delta}{1+\Delta B^2(z)}.$$

It is easily seen that $g \in H_\infty$ and $\|I_\tau g\|_{l_p^n} = \delta$. Thus,

$$\begin{aligned} e_{np}(H_\infty(\mathcal{E}_c),\delta) &\ge \inf_{z_j\in(-1,1)}\left|\int_{-\sqrt{k}}^{\sqrt{k}} g(z)q(z)\,dz\right| \\ &= \inf_{z_j\in(-1,1)}\left|\int_{-\sqrt{k}}^{\sqrt{k}} \varphi\left(\frac{4}{\pi}\arctan B(z)\right) q(z)\,dz\right|, \end{aligned}$$

where the function φ is defined by (5.20). By Theorem 2.15 we get

$$e_{np}(H_\infty(\mathcal{E}_c),\delta) \ge \frac{\pi}{\Lambda}J_2(\lambda,\Delta).$$

It follows from the same theorem that the nodes (2.31) of extremal functions are unique. By conformal mapping of the unit disk onto the interior of the ellipse $\mathcal{E}_c$ these nodes transform to the Chebyshev nodes. □

5.3 Estimates of Optimal Quadratures

We consider now optimal quadrature formulas for the class H_p, $(a,b) = (-1,1)$, and $p(x)\equiv 1$. In view of (5.2) we have

$$\begin{aligned} e(\nu,H_p) &= \inf_{-1<x_1<\dots<x_n<1} \sup_{\substack{f\in H_p \\ f^{(m)}(x_j)=0,\ j=1,\dots,n,\ m=0,\dots,\nu_j-1}} \left|\int_{-1}^{1} f(x)\,dx\right| \\ &= \inf_{-1<x_1<\dots<x_n<1} \sup_{g\in H_p}\left|\int_{-1}^{1}\prod_{j=1}^{n}\left(\frac{x-x_j}{1-x_jx}\right)^{\nu_j} g(x)\,dx\right|. \end{aligned}$$

For

$$N = \sum_{j=1}^{n} \nu_j \quad \text{and} \quad t = (t_1, \dots, t_N) \in I^N := [-1, 1]^N$$

we set

$$B_t(z) := \prod_{j=1}^{N} \frac{z - t_j}{1 - t_j z}, \quad H_p(t) := \{ B_t g : g \in H_p \},$$

$$E_p(t) = \sup_{f \in H_p(t)} \left| \int_{-1}^{1} f(x)\, dx \right|.$$

Denote by $\Omega(\nu)$ the set of all systems of points (5.1) with $-1 < x_1 < \dots < x_n < 1$. If $t \in \Omega(\nu)$ we identify it with

$$(\underbrace{t_1, \dots, t_1}_{\nu_1}, \dots, \underbrace{t_n, \dots, t_n}_{\nu_n}).$$

Then

$$e(\nu, H_p) = \inf_{t \in \Omega(\nu)} E_p(t).$$

Denote by $\overline{\Omega}$ the closure of $\Omega(\nu)$ in the case when $\nu_1, \dots, \nu_n$ are all even.

Proposition 5.8. *Let $1 < p \le \infty$ and $\nu_1, \dots, \nu_n$ be even. Then*

(i) for every $t \in \overline{\Omega}$ there exists a unique function $f_t \in H_p(t)$ such that

$$E_p(t) = \int_{-1}^{1} f_t(x)\, dx,$$

(ii) for every $t \in \overline{\Omega}$

$$g_t(x) = \frac{f_t(x)}{B_t(x)} > 0$$

for $-1 < x < 1$. In particular, if

$$\tau_\nu := \begin{pmatrix} x_1 & \dots & x_n \\ \nu_1 & \dots & \nu_n \end{pmatrix},$$

then

$$f_{\tau_\nu}^{(\nu_k)}(x_k) > 0, \quad k = 1, \dots, n.$$

(iii) $E_p(t)$ is continuous on $\overline{\Omega}$,

(iv) *if $t \in \overline{\Omega}$ and $t \to x \in \overline{\Omega}$, then for all $m \geq 0$, $f_t^{(m)}(z) \to f_x^{(m)}(z)$ uniformly on every compact subset of the unit disk D.*

Proof. Parts (i) and (ii) follow from Lemma 5.1. To prove (iii) we consider the continuous linear functional L_t on $\mathcal{H}_p$ defined by

$$L_t(g) := \int_{-1}^{1} B_t(x)g(x)\,dx.$$

From the definition of g_t we have

$$L_t(g_t) = \sup_{g \in H_p} |L_t(g)| = E_p(t).$$

If $x \in \overline{\Omega}$ and $t \to x \in \overline{\Omega}$, then $\|B_t - B_x\|_{L_q(-1,1)} \to 0$ for all $q < \infty$. This implies that $\|L_t - L_x\| \to 0$ from which immediately follows that $E_p(t) \to E_p(x)$. Thus (iii) is proved.

Let $1 < p < \infty$. For $t \to x$ we have

$$L_t(g_t) = L_t(g_t) + (L_x - L_t)(g_t) \to E_p(x).$$

Consequently, any weakly convergent subsequence of $\{g_t\}$ must converge to a function $g \in H_p$ for which $L_x(g) = E_p(x)$. According to (i) only g_x has this property. Since H_p is sequentially weak compact, we get $g_t \to g_x$ weakly. But $\|g_t\|_{\mathcal{H}_p} = 1 = \|g_x\|_{\mathcal{H}_p}$. So for $1 < p < \infty$ we also have $\|g_t - g_x\|_{\mathcal{H}_p} \to 0$. It is easy to see that $\|f_t - f_x\|_{\mathcal{H}_p} \to 0$ if and only if $\|g_t - g_x\|_{\mathcal{H}_p} \to 0$ where $f_t = g_t B_t$ and $f_x = g_x B_x$. Now (iv) is an immediate consequence of Cauchy's formula

$$f_t^{(m)}(z) = \frac{m!}{2\pi i}\int_{|\xi|=1} \frac{f_t(\xi)}{(\xi - z)^{m+1}}\,d\xi.$$

The case $p = \infty$ easily follows from the fact that for every $t \in \overline{\Omega}$, $g_t(z) \equiv 1$. □

We need also the following auxiliary lemma.

Lemma 5.9. *Let $h > 0$ and $f \in C^r[x_0 - h, x_0 + h]$. Suppose that f has exactly r zeros in $[x_0 - h, x_0 + h]$ and $f(x_0 - h) = f(x_0 + h) = 0$. If $0 < m < f^{(r)}(t) < M$ for all $t \in [x_0 - h, x_0 + h]$, then*

$$t_2 - t_1 > 4\sqrt{\frac{m}{Mr!\sqrt{e}\,2^{r-1}}}\,h,$$

where $t_1 < t_2$ are the zeros of $f^{(r-2)}(t)$ in $[x_0 - h, x_0 + h]$.

Proof. By Rolle's theorem $f^{(k)}(t)$ has exactly $r-k$ zeros in $[x_0-h, x_0+h]$. Denote them by $t_{k1} \leq \ldots \leq t_{k,r-k}$. Evidently for $k \leq r-2$

$$f^{(k)}(x) = \int_{t_{k2}}^{x} f^{(k+1)}(t)\,dt.$$

Thus for $k \leq r-3$

$$\begin{aligned}\max_{t_{k1}\leq x\leq t_{k,r-k}} |f^{(k)}(x)| &= \max_{t_{k+1,1}\leq x\leq t_{k+1,r-k-1}} |f^{(k)}(x)| \\ &\leq (t_{k+1,r-k+1} - t_{k+1,1}) \max_{t_{k+1,1}\leq x\leq t_{k+1,r-k-1}} |f^{(k+1)}(x)| \\ &\leq 2h \max_{t_{k+1,1}\leq x\leq t_{k+1,r-k-1}} |f^{(k+1)}(x)|. \qquad (5.23)\end{aligned}$$

From Newton's interpolation formula it follows that for all $x \in [t_1, t_2]$

$$f^{(r-2)}(x) = (x-t_1)(x-t_2)\frac{f^{(r)}(\zeta)}{2}, \quad t_1 < \zeta < t_2.$$

Consequently,

$$\max_{t_1\leq x\leq t_2} |f^{(r-2)}(x)| \leq \frac{(t_2-t_1)^2}{8}M.$$

Set $\xi = t_{r-1,1}$. Then repeated use of (5.23) gives

$$|f^{(k)}(\xi)| \leq (2h)^{r-k-2}\frac{(t_2-t_1)^2}{8}M. \tag{5.24}$$

Now suppose that $\xi \leq x_0$. By Taylor's formula

$$f(x) = \sum_{k=0}^{r-1} \frac{f^{(k)}(\xi)}{k!}(x-\xi)^k + \frac{f^{(r)}(x+\theta(x-\xi))}{r!}(x-\xi)^r, \quad 0<\theta<1.$$

Using (5.24) and the fact that $f^{(r-1)}(\xi) = 0$, we get for $x \geq x_0$

$$\begin{aligned} f(x) &\geq \frac{m}{r!}(x-x_0)^r - \sum_{k=0}^{r-2} \frac{|f^{(k)}(\xi)|}{k!}(x-\xi)^k \\ &\geq \frac{m}{r!}(x-x_0)^r - \frac{(t_2-t_1)^2}{8}M\sum_{k=0}^{r-2}\frac{(2h)^{r-k-2}}{k!}h^k \\ &> \frac{m}{r!}(x-x_0)^r - \frac{(t_2-t_1)^2}{8}M(2h)^{r-2}\sqrt{e}.\end{aligned}$$

Substituting $x = x_0 + h$ we obtain

$$0 > \frac{m}{r!}h^r - \frac{(t_2 - t_1)^2}{8} M(2h)^{r-2}\sqrt{e}.$$

From the last inequality our assertion follows immediately.

Suppose that $x_0 \leq \xi$. Let $x \leq x_0$. In a similar fashion we get

$$f(x) < -\frac{m}{r!}(x_0 - x)^r + \frac{(t_2 - t_1)^2}{8} M(2h)^{r-2}\sqrt{e}$$

for odd r, and

$$f(x) > \frac{m}{r!}(x_0 - x)^r - \frac{(t_2 - t_1)^2}{8} M(2h)^{r-2}\sqrt{e}$$

for even r. Putting $x = x_0 - h$ and taking into account that $f(x_0 - h) = 0$, we obtain the desired inequality. □

Theorem 5.10. *For every system of natural numbers $\nu_1, \ldots, \nu_n$ there exist optimal nodes of type $\nu = (\nu_1, \ldots, \nu_n)$ in the class H_p, $1 < p \leq \infty$, such that $-1 < x_1, \ldots, x_n < 1$. Moreover, the coefficients of the optimal quadrature formula satisfy the relations*

$$\begin{aligned} &a_{k,\nu_k-1} = 0, \quad a_{k,\nu_k-2} > 0, \quad \text{if } \nu_k \text{ is even,} \\ &a_{k,\nu_k-1} > 0, \qquad\qquad\qquad\quad \text{if } \nu_k \text{ is odd.} \end{aligned} \tag{5.25}$$

Proof. Assume first that $\nu_1, \ldots, \nu_n$ are even. Since $E_p(t)$ is continuous for $t \in \overline{\Omega}$ it has a minimum on $\overline{\Omega}$. The points in $\overline{\Omega}$ have representations of the form

$$\tau_\mu = \begin{pmatrix} -1 & x_1 & \ldots & x_m & 1 \\ \mu_0 & \mu_1 & \ldots & \mu_m & \mu_{m+1} \end{pmatrix},$$

where $m \leq n$, $-1 < x_1 < \ldots < x_m < 1$, and the μ_k's are the corresponding multiplicities. We always include ± 1 for technical reasons. Therefore we allow the possibility that $\mu_0 = 0$ or $\mu_{m+1} = 0$. With this exception the μ_k's are sums of consecutive ν_i's.

We now claim that the minimum of $E_p(t)$ on $\overline{\Omega}$ is obtained at a point τ_μ with $\mu_0 = \mu_{m+1} = 0$. Assume that $\mu_0 + \mu_{m+1} > 0$. Consider the point in $\overline{\Omega}$

$$\tau'_\mu = \begin{pmatrix} x_1 & x_2 & \ldots & x_{m-1} & x_m \\ \mu_0 + \mu_1 & \mu_2 & \ldots & \mu_{m-1} & \mu_m + \mu_{m+1} \end{pmatrix}.$$

Since all the μ_k's are even, we have on $(-1, 1) \setminus \{x_1, \ldots, x_m\}$ that

$$B_{\tau'_\mu}(x) < B_{\tau_\mu}(x).$$

Using Proposition 5.8, we obtain

$$E_p(\tau'_\mu) = \int_{-1}^{1} g_{\tau'_\mu}(x) B_{\tau'_\mu}(x)\, dx < \int_{-1}^{1} g_{\tau'_\mu}(x) B_{\tau_\mu}(x)\, dx$$
$$\leq \int_{-1}^{1} g_{\tau_\mu}(x) B_{\tau_\mu}(x)\, dx = E_p(\tau_\mu).$$

Thereby our claim is justified.

Let us suppose that the nodes

$$\tau_\mu = \begin{pmatrix} x_1 & \dots & x_m \\ \mu_1 & \dots & \mu_m \end{pmatrix} \tag{5.26}$$

are optimal. Let k be fixed. For sufficiently small $|h|$ we define the nodes

$$\tau_\mu(h) = \begin{pmatrix} x_1 & \dots & x_{k-1} & x_k + h & x_{k+1} & \dots & x_m \\ \mu_1 & \dots & \mu_{k-1} & \mu_k & \mu_{k+1} & \dots & \mu_m \end{pmatrix}.$$

Since τ_μ is optimal,

$$E_p(\tau_\mu) \leq E_p(\tau_\mu(h)). \tag{5.27}$$

We have

$$\int_{-1}^{1} f_{\tau_\mu(h)}(x)\, dx - \sum_{j=1}^{m} \sum_{s=0}^{\mu_j - 1} a_{js} f^{(s)}_{\tau_\mu(h)}(x_j) \leq E_p(\tau_\mu),$$

where a_{js} are optimal coefficients for the nodes τ_μ. Taking into account properties of $f_{\tau_\mu(h)}$, we obtain

$$E_p(\tau_\mu(h)) - \sum_{s=0}^{\mu_k - 1} a_{js} f^{(s)}_{\tau_\mu(h)}(x_k) \leq E_p(\tau_\mu). \tag{5.28}$$

By Taylor's formula

$$f^{(s)}_{\tau_\mu(h)}(x_k) = \frac{f^{(\mu_k)}_{\tau_\mu(h)}(x_k + \theta h)}{(\mu_k - s)!}(-h)^{\mu_k - s}, \quad 0 < \theta < 1. \tag{5.29}$$

In view of Proposition 5.8 there exist constants $C_1 > 0$ and $C_2 > 0$ such that

$$C_1 \leq f^{(\mu_k)}_{\tau_\mu(h)}(x_k + h) \leq C_2$$

for all sufficiently small $|h|$. Thus

$$f_{\tau_\mu(h)}^{(s)}(x_k) = O\left(h^{\mu_j - s}\right), \quad s = 0, \dots, \mu_j - 2,$$

and

$$f_{\tau_\mu(h)}^{(\mu_j - 1)}(x_k) = -C(h)h,$$

where $C_1 \leq C(h) \leq C_2$.

It follows from (5.28) that

$$E_p(\tau_\mu(h)) + O(h^2) + a_{k,\mu_k-1}C(h)h \leq E_p(\tau_\mu).$$

This inequality contradicts (5.27) in the case $a_{k,\mu_k-1} \neq 0$. Hence $a_{k,\mu_k-1} = 0$. Now (5.28) implies

$$E_p(\tau_\mu(h)) + O(h^3) - a_{k,\mu_k-2}f_{\tau_\mu(h)}^{(\mu_k-2)}(x_k) \leq E_p(\tau_\mu).$$

But in view of (5.29)

$$f_{\tau_\mu(h)}^{(\mu_k-2)}(x_k) \geq \frac{C_1}{2}h^2.$$

Thus (5.27) yields $a_{k,\mu_k-2} \geq 0$. If $a_{k,\mu_k-2} = 0$, then $E_p(\tau'_\mu) = E_p(\tau_\mu)$, where

$$\tau'_\mu = \begin{pmatrix} x_1 & \dots & x_{k-1} & x_k & x_{k+1} & \dots & x_m \\ \mu_1 & \dots & \mu_{k-1} & \mu_k - 2 & \mu_{k+1} & \dots & \mu_m \end{pmatrix}.$$

Consequently, $f_{\tau_\mu} = f_{\tau'_\mu}$ and $f_{\tau'_\mu}^{(\mu_k-2)}(x_k) = 0$ which contradicts (ii) of Proposition 5.8.

Suppose that

$$\inf_{\tau_\nu \in \Omega(\nu)} E_p(\tau_\nu) = E_p(\tau_\mu),$$

where τ_μ is defined by (5.26). We have to prove that $m = n$ which will imply that $\mu_k = \nu_k$, $k = 1, \dots, n$. If $m < n$, then there is an index k such that $\mu_k = \nu_k + \dots + \nu_{k+j}$ for some $j > 0$. For each sufficiently small $h > 0$ we define the nodes

$$\tau'_\mu(h) = \begin{pmatrix} x_1 & \dots & x_{k-1} & \varepsilon - h & \varepsilon + h & x_{k+1} & \dots & x_m \\ \mu_1 & \dots & \mu_{k-1} & \nu_k & \mu_k - \nu_k & \mu_{k+1} & \dots & \mu_m \end{pmatrix},$$

where $\varepsilon \in [x_k - h, x_k + h]$. Let us show that there exists such ε for which

$$f_{\tau'_\mu(h)}^{(\mu_k-1)}(x_k) = 0. \tag{5.30}$$

It follows from Proposition 5.8 that for sufficiently small h

$$f^{(\mu_k)}_{\tau'_\mu(h)}(t) \geq C > 0$$

for all $t \in [x_k - 2h, x_k + 2h]$. According to Rolle's theorem $f^{(\lambda)}_{\tau'_\mu(h)}(t)$, $\lambda = 0, \dots, \mu_k - 1$, has exactly $\mu_k - \lambda$ zeros in $[x_k - 2h, x_k + 2h]$. It is easy to see that the unique zero $\xi(\varepsilon)$ of $f^{(\mu_k-1)}_{\tau'_\mu(h)}(t)$ for $t \in [x_k - 2h, x_k + 2h]$ is a continuous function of ε for fixed h. Moreover, $\xi(x_k - h) < x_k$ and $\xi(x_k + h) > x_k$. Consequently, there exists $\varepsilon \in [x_k - h, x_k + h]$ for which (5.30) holds. In what follows we assume that ε is chosen satisfying (5.30).

Denote by $t_{\lambda 1} \leq \dots \leq t_{\lambda,\mu_k-\lambda}$ the zeros of $f^{(\lambda)}_{\tau'_\mu(h)}(t)$ in $[x_k - 2h, x_k + 2h]$. By Newton's interpolation formula for every $t \in [x_k - 2h, x_k + 2h]$

$$f^{(\lambda)}_{\tau'_\mu(h)}(t) = (t - t_{\lambda 1}) \dots (t - t_{\lambda,\mu_k-\lambda}) \frac{f^{(\mu_k)}_{\tau'_\mu(h)}(\zeta)}{(\mu_k - \lambda)!}, \tag{5.31}$$

where $x_k - 2h < \zeta < x_k + 2h$. Hence there exists a positive constant A such that

$$|f^{(\lambda)}_{\tau'_\mu(h)}(x_k)| \leq A h^{\mu_k - \lambda}, \quad \lambda = 0, \dots, \mu_k - 1. \tag{5.32}$$

Since $t_{\mu_k-1,1} = x_k$, using (5.31), we have

$$|f^{(\mu_k-2)}_{\tau'_\mu(h)}(x_k)| = \max_{t \in [t_{\mu_k-2,1}, t_{\mu_k-2,2}]} |f^{(\mu_k-2)}_{\tau'_\mu(h)}(t)| \geq \frac{C_1}{8}(t_{\mu_k-2,2} - t_{\mu_k-2,1})^2.$$

According to Lemma 5.9, there is a constant $A_1 > 0$ such that

$$|f^{(\mu_k-2)}_{\tau'_\mu(h)}(x_k)| > A_1 h^2. \tag{5.33}$$

From (5.31) we see that

$$f^{(\mu_k-2)}_{\tau'_\mu(h)}(x_k) < 0. \tag{5.34}$$

We have

$$E_p(\tau'_\mu(h)) - \sum_{j=1}^{m} \sum_{s=0}^{\mu_j - 1} a_{js} f^{(s)}_{\tau'_\mu(h)}(x_j) \leq E_p(\tau_\mu).$$

Taking into account (5.32)-(5.34) and the fact that $a_{k,\mu_k-2} > 0$, we obtain

$$\begin{aligned} E_p(\tau_\mu) &\geq E_p(\tau'_\mu(h)) + O(h^3) - a_{k,\mu_k-2} f^{(\mu_k-2)}_{\tau'_\mu(h)}(x_k) \\ &> E_p(\tau'_\mu(h)) + O(h^3) + a_{k,\mu_k-2} A_1 h^2. \end{aligned}$$

This contradicts the assumption that τ_μ is optimal. Therefore $m = n$ and the theorem is proved in the case of even ν_k's.

Now let $\nu_1, \dots, \nu_n$ be arbitrary natural numbers. Set

$$\mu_k := 2\left[\frac{\nu_k + 1}{2}\right], \quad k = 1, \dots, n.$$

Then there exist optimal nodes

$$\tau_\mu = \begin{pmatrix} x_1 & \dots & x_n \\ \mu_1 & \dots & \mu_n \end{pmatrix}.$$

Since the coefficients of the optimal quadrature formula for these nodes satisfy the relations

$$a_{k,\mu_k - 1} = 0, \quad k = 1, \dots, n,$$

the nodes

$$\tau_\nu = \begin{pmatrix} x_1 & \dots & x_n \\ \nu_1 & \dots & \nu_n \end{pmatrix}$$

are optimal of the type $\nu = (\nu_1, \dots, \nu_n)$. □

Now we can obtain estimates for $e(\nu, H_p)$.

Theorem 5.11. *Let $s \in N$ and $1 \le p \le \infty$. Then there exist constants C_1 and C_2 such that for all $2s - 1 \le \nu_j \le 2s$, $j = 1, \dots, n$,*

$$C_1 n^{1/(2q)} \exp\left(-\pi\sqrt{\frac{ns}{q}}\right) \le e(\nu, H_p) \le C_2 n^{1/(2q)} \exp\left(-\pi\sqrt{\frac{ns}{q}}\right),$$

were $1/p + 1/q = 1$.

Proof. It follows from Theorem 5.10 that for all $2s - 1 \le \nu_j \le 2s$, $j = 1, \dots, n$,

$$e(\nu, H_p) = e(\nu_s, H_p),$$

where $\nu_s = (2s, \dots, 2s)$. Consequently,

$$e(\nu, H_p) = \inf_{-1 < x_1 < \dots < x_n < 1} \sup_{g \in H_p} \left| \int_{-1}^{1} g(x) B^{2s}(x)\, dx \right|,$$

where

$$B(x) = \prod_{j=1}^{n} \frac{x - x_j}{1 - x_j x}.$$

From Lemma 5.1 we have

$$\sup_{g\in H_p}\left|\int_{-1}^{1} g(x)B^{2s}(x)\,dx\right| = \sup_{g\in H_p}\int_{-1}^{1} |g(x)|B^{2s}(x)\,dx.$$

In view of (2.32) the infimum in the definition of $E_n(p,q,m)$ (see (4.10)) can be considered over the Blaschke products with real zeros. Hence

$$e(\nu, H_p) = E_n(p, 1, 2s).$$

It remains to apply Theorem 4.11. □

5.4 Notes and References

5.1. For the class $H_{\infty,R}$ and $\nu_1 = \ldots = \nu_n = 2$ the best quadrature formula was constructed by Bojanov [1974]. For $p = \infty$ Theorem 5.2 was proved in Osipenko [1988].

5.2. The material of this section is taken from Osipenko [1995a], [1994c]. For the class $H_\infty(\mathcal{E}_c)$ estimates of optimal quadrature formula were obtained by Bakhvalov [1967]. The problem of the existence and uniqueness of optimal nodes for classes of analytic functions was discussed in Barrar, Loeb, Werner [1974], Loeb [1974], Bojanov [1979], and Osipenko [1988].

5.3. Proposition 5.8 is due to Anderson, Bojanov [1984]. Lemma 5.9 is taken from Bojanov [1978] (we give a little more precise inequality). Theorem 5.10 was proved by Anderson, Bojanov [1984] (see also Bojanov [1979]). The estimates for $e(\nu, H_p)$ were studied by many authors (Bojanov [1973], Loeb, Werner [1974], Stenger [1978], Newman [1979], Anderson [1980]). The most precise result was obtained by Anderson, Bojanov [1984] (Theorem 5.11).

Appendix. Elliptic Integrals and Elliptic Functions

In this appendix we shall list some facts about elliptic integrals and elliptic functions which we used in the main text. More detailed information may be found, for example, in Akhiezer [1970].

We begin with elliptic integrals. The integrals

$$\begin{aligned} K &:= \int_0^1 \frac{dt}{\sqrt{(1-t^2)(1-k^2t^2)}}, \\ K' &:= \int_0^1 \frac{dt}{\sqrt{(1-t^2)(1-k'^2t^2)}} \end{aligned} \tag{A.35}$$

are called the *complete elliptic integrals of the first kind* for the moduli k and $k' = \sqrt{1-k^2}$. For our purpose it suffices to deal with the so-called normal case when $0 < k < 1$.

The integral K may be represented by the series

$$K = \frac{\pi}{2} \sum_{m=0}^{\infty} \frac{((2m)!)^2}{2^{4m}(m!)^4} k^{2m}.$$

Thus for $k \to 0$ we have

$$K = \frac{\pi}{2} + O(k^2).$$

Set

$$h := e^{-\frac{\pi K'}{K}}.$$

Then the following equalities

$$K = \frac{\pi}{2}\left(1 + 2\sum_{m=1}^{\infty} h^{m^2}\right)^2, \tag{A.36}$$

$$\sqrt{k} = 2h^{1/4}\frac{\sum_{m=0}^{\infty} h^{m(m+1)}}{1 + 2\sum_{m=1}^{\infty} h^{m^2}} \tag{A.37}$$

hold. Consequently, k satisfies the equality

$$\frac{\pi K'}{2K} = s$$

if and only if $k = \kappa(s)$ where

$$\kappa(s) = 4e^{-s}\left(\frac{\sum_{m=0}^{\infty} e^{-2sm(m+1)}}{1 + 2\sum_{m=1}^{\infty} e^{-2sm^2}}\right)^2. \tag{A.38}$$

Hence

$$\kappa(s) = 4e^{-s} + O\left(e^{-3s}\right). \tag{A.39}$$

From (A.37) it follows that

$$\sqrt{k} = 2h^{1/4} + O\left(h^{5/4}\right) \tag{A.40}$$

as $h \to 0$ and

$$h = \frac{1}{16}k^2 + O\left(k^4\right)$$

as $k \to 0$. Thus taking into account (A.36), we have

$$K' = -\frac{K}{\pi}\log h = \log\frac{4}{k} + O\left(k^2 \log\frac{4}{k}\right). \tag{A.41}$$

Set

$$H_0 := \prod_{m=1}^{\infty}(1 - h^{2m}).$$

The *theta functions* are defined as follows

$$\theta_1(u) := 2H_0 h^{1/4} \sin\frac{\pi u}{2K} \prod_{m=1}^{\infty} \left(1 - 2h^{2m}\cos\frac{\pi u}{K} + h^{4m}\right),$$

$$\theta_2(u) := 2H_0 h^{1/4} \cos\frac{\pi u}{2K} \prod_{m=1}^{\infty} \left(1 + 2h^{2m}\cos\frac{\pi u}{K} + h^{4m}\right),$$

$$\theta_3(u) := H_0 \prod_{m=1}^{\infty} \left(1 + 2h^{2m-1}\cos\frac{\pi u}{K} + h^{4m-2}\right),$$

$$\theta_4(u) := H_0 \prod_{m=1}^{\infty} \left(1 - 2h^{2m-1}\cos\frac{\pi u}{K} + h^{4m-2}\right).$$

The functions

$$\begin{aligned} \operatorname{sn}(u,k) &:= \frac{1}{\sqrt{k}}\frac{\theta_1(u)}{\theta_0(u)}, \\ \operatorname{cn}(u,k) &:= \sqrt{\frac{k'}{k}}\frac{\theta_2(u)}{\theta_0(u)}, \\ \operatorname{dn}(u,k) &:= \sqrt{k'}\frac{\theta_3(u)}{\theta_0(u)} \end{aligned} \tag{A.42}$$

are called the *Jacobi functions.* As usual we do not indicate the dependence of these functions on the modulus k. For the Jacobi functions the following equalities

$$\operatorname{cn}^2 u = 1 - \operatorname{sn}^2 u, \quad \operatorname{dn}^2 u = 1 - k^2 \operatorname{sn}^2 u \tag{A.43}$$

hold.

Zeros, poles, and periods of the Jacobi functions are given in the table

	Zeros	Poles	Periods
$\operatorname{sn} u$	$2nK + 2mK'$	$2nK + (2m+1)K'$	$4K,\ 2K'$
$\operatorname{cn} u$	$(2n+1)K + 2mK'$	$2nK + (2m+1)K'$	$4K,\ 2K + 2K'$
$\operatorname{dn} u$	$(2n+1)K + (2m+1)K'$	$2nK + (2m+1)K'$	$2K,\ 4K'$

We give one more table connected with some transformations of the Jacobi functions

	$u+K$	$u+2K$
$\operatorname{sn} u$	$\dfrac{\operatorname{cn} u}{\operatorname{dn} u}$	$-\operatorname{sn} u$
$\operatorname{cn} u$	$-k'\dfrac{\operatorname{sn} u}{\operatorname{dn} u}$	$-\operatorname{cn} u$
$\operatorname{dn} u$	$k'\dfrac{1}{\operatorname{dn} u}$	$\operatorname{dn} u$

The summation formulas for the Jacobi functions are given by

$$\begin{aligned}
\operatorname{sn}(u+v) &= \frac{\operatorname{sn} u \operatorname{cn} v \operatorname{dn} v + \operatorname{sn} v \operatorname{cn} u \operatorname{dn} u}{1-k^2 \operatorname{sn}^2 u \operatorname{sn}^2 v},\\
\operatorname{cn}(u+v) &= \frac{\operatorname{cn} u \operatorname{cn} v - \operatorname{sn} u \operatorname{dn} u \operatorname{sn} v \operatorname{dn} v}{1-k^2 \operatorname{sn}^2 u \operatorname{sn}^2 v},\\
\operatorname{dn}(u+v) &= \frac{\operatorname{dn} u \operatorname{dn} v - k^2 \operatorname{sn} u \operatorname{cn} u \operatorname{sn} v \operatorname{cn} v}{1-k^2 \operatorname{sn}^2 u \operatorname{sn}^2 v}.
\end{aligned}$$

From the first formula and (A.43) follows that

$$\operatorname{sn}(u+v)\operatorname{sn}(u-v) = \frac{\operatorname{sn}^2 u - \operatorname{sn}^2 v}{1-k^2 \operatorname{sn}^2 u \operatorname{sn}^2 v}.$$

We list some particular values of the Jacobi functions

$$\begin{aligned}
&\operatorname{sn} 0 = 0, && \operatorname{cn} 0 = 1, && \operatorname{dn} 0 = 1,\\
&\operatorname{sn} K = 1, && \operatorname{cn} K = 0, && \operatorname{dn} K = k',\\
&\operatorname{sn}\frac{K}{2} = \frac{1}{\sqrt{1+k'}}, && \operatorname{cn}\frac{K}{2} = \frac{\sqrt{k'}}{\sqrt{1+k'}}, && \operatorname{dn}\frac{K}{2} = \sqrt{k'},\\
&\operatorname{sn}\frac{iK'}{2} = \frac{i}{\sqrt{k}}, && \operatorname{cn}\frac{iK'}{2} = \frac{\sqrt{1+k}}{\sqrt{k}}, && \operatorname{dn}\frac{iK'}{2} = \sqrt{1+k}.
\end{aligned}$$

From these equalities and summation formulas we have that for all real u

$$\sqrt{k}\left|\operatorname{sn}\left(u+\frac{iK'}{2}\right)\right| \equiv 1. \tag{A.44}$$

In what follows we denote by Λ an Λ' the complete elliptic integrals of the first kind for the moduli λ and $\lambda' = \sqrt{1-\lambda^2}$. The *Landen transform* is given by

$$\lambda = \frac{1-k'}{1+k'}, \qquad M = \frac{1}{1+k'},$$
$$\Lambda = \frac{K}{2M}, \qquad \Lambda' = \frac{K'}{M},$$

$$\begin{aligned}
\operatorname{sn}\left(\frac{u}{M}, \lambda\right) &= \frac{1}{M}\frac{\operatorname{sn}(u,k)\operatorname{cn}(u,k)}{\operatorname{dn}(u,k)},\\
\operatorname{cn}\left(\frac{u}{M}, \lambda\right) &= \frac{1-(1+k')\operatorname{sn}^2(u,k)}{\operatorname{dn}(u,k)},\\
\operatorname{dn}\left(\frac{u}{M}, \lambda\right) &= \frac{1-(1-k')\operatorname{sn}^2(u,k)}{\operatorname{dn}(u,k)}.
\end{aligned}$$

The *Gauss transform* is given by

$$\lambda = \frac{2\sqrt{k}}{1+k}, \qquad M = \frac{1}{1+k},$$
$$\Lambda = \frac{K}{M}, \qquad \Lambda' = \frac{K'}{2M},$$

$$\begin{aligned}
\operatorname{sn}\left(\frac{u}{M}, \lambda\right) &= \frac{1}{M}\frac{\operatorname{sn}(u,k)}{1+k\operatorname{sn}^2(u,k)},\\
\operatorname{cn}\left(\frac{u}{M}, \lambda\right) &= \frac{\operatorname{cn}(u,k)\operatorname{dn}(u,k)}{1+k\operatorname{sn}^2(u,k)},\\
\operatorname{dn}\left(\frac{u}{M}, \lambda\right) &= \frac{1-k\operatorname{sn}^2(u,k)}{1+k\operatorname{sn}^2(u,k)}.
\end{aligned}$$

The *first principal transform of degree n* is defined by the following equalities

$$\Lambda = \frac{K}{nM}, \quad \Lambda' = K'M,$$
$$c_r = \operatorname{sn}^2\left(\frac{rK}{n}, k\right),$$
$$\lambda = k^n \prod_{r=1}^{[n/2]} c_{2r-1}^2, \quad M = \prod_{r=1}^{[n/2]} \frac{c_{2r-1}}{c_{2r}}.$$

If n is odd, then

$$\operatorname{sn}\left(\frac{u}{M}, \lambda\right) = \frac{1}{M}\operatorname{sn}(u,k) \prod_{r=1}^{(n-1)/2} \frac{1 - \dfrac{\operatorname{sn}^2(u,k)}{c_{2r}}}{1 - k^2 c_{2r}\operatorname{sn}^2(u,k)},$$
$$\operatorname{cn}\left(\frac{u}{M}, \lambda\right) = \operatorname{cn}(u,k) \prod_{r=1}^{(n-1)/2} \frac{1 - \dfrac{\operatorname{sn}^2(u,k)}{c_{2r-1}}}{1 - k^2 c_{2r}\operatorname{sn}^2(u,k)},$$
$$\operatorname{dn}\left(\frac{u}{M}, \lambda\right) = \operatorname{dn}(u,k) \prod_{r=1}^{(n-1)/2} \frac{1 - k^2 c_{2r-1}\operatorname{sn}^2(u,k)}{1 - k^2 c_{2r}\operatorname{sn}^2(u,k)}.$$

If n is even, then

$$\operatorname{sn}\left(\frac{u}{M} + \Lambda, \lambda\right) = \prod_{r=1}^{n/2} \frac{1 - \dfrac{\operatorname{sn}^2(u,k)}{c_{2r-1}}}{1 - k^2 c_{2r-1}\operatorname{sn}^2(u,k)},$$
$$\operatorname{cn}\left(\frac{u}{M} + \Lambda, \lambda\right) = -\frac{\lambda' \operatorname{sn}(u,k)}{M \operatorname{cn}(u,k)} \prod_{r=1}^{n/2} \frac{1 - \dfrac{\operatorname{sn}^2(u,k)}{c_{2r}}}{1 - k^2 c_{2r-1}\operatorname{sn}^2(u,k)},$$
$$\operatorname{dn}\left(\frac{u}{M} + \Lambda, \lambda\right) = \frac{\lambda'}{\operatorname{dn}(u,k)} \prod_{r=1}^{n/2} \frac{1 - k^2 c_{2r}\operatorname{sn}^2(u,k)}{1 - k^2 c_{2r-1}\operatorname{sn}^2(u,k)}.$$

The *second principal transform of the first degree* is defined as follows

$$\lambda = k', \qquad M = \frac{1}{i},$$
$$\Lambda = \frac{iK'}{M}, \qquad i\Lambda' = -\frac{K}{M},$$

$$\operatorname{sn}\left(\frac{u}{M},\lambda\right)=\frac{1}{M}\frac{\operatorname{sn}(u,k)}{\operatorname{cn}(u,k)},$$
$$\operatorname{cn}\left(\frac{u}{M},\lambda\right)=\frac{1}{\operatorname{cn}(u,k)},$$
$$\operatorname{dn}\left(\frac{u}{M},\lambda\right)=\frac{\operatorname{dn}(u,k)}{\operatorname{cn}(u,k)}.$$

The derivatives of the Jacobi functions satisfy the equalities

$$\frac{d}{du}\operatorname{sn}u=\operatorname{cn}u\operatorname{dn}u,$$
$$\frac{d}{du}\operatorname{cn}u=-\operatorname{sn}u\operatorname{dn}u,$$
$$\frac{d}{du}\operatorname{dn}u=-k^2\operatorname{sn}u\operatorname{cn}u.$$

Thus we have

$$\begin{aligned}\operatorname{sn}u&=u-\frac{1+k^2}{6}u^3+O(u^5),\\ \operatorname{cn}u&=1-\frac{1}{2}u^2+O(u^4),\\ \operatorname{dn}u&=1-\frac{k^2}{2}u^2+O(u^4).\end{aligned}\tag{A.45}$$

For the integrals of the Jacobi functions we have

$$\begin{aligned}\int_0^z\operatorname{sn}u\,du&=-\frac{1}{k}\log(\operatorname{dn}z+k\operatorname{cn}z)+\frac{1}{k}\log(1+k),\\ \int_0^z\operatorname{cn}u\,du&=\frac{i}{k}\log(\operatorname{dn}z-ik\operatorname{sn}z),\\ \int_0^z\operatorname{dn}u\,du&=i\log(\operatorname{cn}z-i\operatorname{sn}z).\end{aligned}\tag{A.46}$$

The expansion of the Jacobi functions in a Fourier series has the following form:

$$\begin{aligned}\operatorname{sn}\left(\frac{2Ku}{\pi},k\right)&=\frac{2\pi}{kK}\sum_{n=1}^{\infty}\frac{h^{n-1/2}}{1-h^{2n-1}}\sin(2n-1)u,\\ \operatorname{cn}\left(\frac{2Ku}{\pi},k\right)&=\frac{2\pi}{kK}\sum_{n=1}^{\infty}\frac{h^{n-1/2}}{1+h^{2n-1}}\cos(2n-1)u,\\ \operatorname{dn}\left(\frac{2Ku}{\pi},k\right)&=\frac{\pi}{2K}+\frac{2\pi}{K}\sum_{n=1}^{\infty}\frac{h^{n}}{1+h^{2n}}\cos 2nu.\end{aligned}\tag{A.47}$$

Bibliography

Arestov, V. V.

[1989] Best recovery of operators and related problems. Trudy Mat. Inst. Steklov. **189**, 3–20. English transl. in Proc. Steklov Inst. Math. **189**, 1991.

Akhiezer, N. I.

[1965] Lectures on the Theory of Approximation, Nauka, Moscow (Russian). English edition: Frederick Ungar, New York, 1956.

[1970] Elements of the Theory of Elliptic Functions, Nauka, Moscow (Russian). English edition: Amer. Math. Soc., Providence, RI, 1990.

Anderson, J.-E.

[1980] Optimal quadrature of H^p functions. Math. Z. **172**, 55–62.

Anderson, J.-E., Bojanov, B. D.

[1984] A note on the optimal quadrature in H_p. Numer. Math. **44**, 301–308.

Babenko, K. I.

[1958] On the best approximation of a class of analytic functions. Izv. Akad. Nauk SSSR. Ser. Mat. **22**, 631–640 (in Russian).

Bakhvalov, N. S.

[1967] On the optimal speed of integrating analytic functions. Zh. Vychisl. Mat. Mat. Fiz. **7**, 1011–1020. English transl. in USSR Comput. Math. Math. Phys. **7** (1967), 63–75.

Barrar, R. B., Loeb, H., Werner, H.

[1974] On the existence of optimal integration formulas for analytic functions. Numer. Math. **23**, 105–117.

Bojanov, B. D.

[1973] Optimal rate of integration and ε-entropy of a class of analytic functions. Mat. Zametki **14**, 3–10. English transl. in Math. Notes **19** (1973), 551–556.

[1974] Best quadrature formula for a certain class of analytic functions. Zastos. Mat. **14**, 441–447.

[1976] Best reconstruction of differentiable periodic functions from their Fourier coefficients. Serdica B'lg. Mat. Spis. **2**, 300–304.

[1978] Existence and characterization of optimal quadrature formulas for a certain class of differentiable functions. J. Approx. Theory **22**, 262–283.

[1979] On the existence of optimal quadrature formulae for smooth functions. Calcolo **16**, 61–70.

Borsuk, K.

[1933] Drei Sätze über die n-dimensionale euklidische Sphäre. Fund. Math. **20**, 177-191.

Burchard, H. G., Höllig, K.

[1985] n-widths and entropy of H^p-spaces in $L_q(-1,1)$. SIAM J. Math. Anal. **16**, 405–421.

Dieudonné, J.

[1931] Recherches sur quelques problems ralatifs aux polinomes et aux fonctions bornées d'une variable complexe. Ann. Ecole Norm. sup. **3**, P.247–358.

Duren, P. L.

[1970] Theory of H_p Spaces. Academic Press, New York.

Duren, P. L., Khavinson, D., Shapiro, H. S., Sundberg, C.

[1993] Contractive zero-divisors in Bergman spaces. Pacific J. Math. **157**, 37–56.

Erokhin, V. P.

[1968] On the best linear approximation of functions analytically extendible from a given continuum to a given domain. Uspekhi Mat. Nauk **23**, 91–132. English transl. in Russian Math. Surveys **23** (1968), 93–135.

Farkov, Yu. A.

[1990] The widths of Hardy and Bergman classes in a ball of C^n. Uspekhi Mat. Nauk **45**, 197–198. English transl. in Russian Math. Surveys **45** (1990), 229–231.

Fisher, S. D.

[1983] Function Theory on Planar Domains: A Second Course in Complex Analysis. Wiley–Interscience, New York.

[1989] Envelopes, widths, and Landau problems for analytic functions. Constr. Approx. **5**, 171–187.

Fisher, S. D., Micchelli, C. A.

[1980] The n-width of sets of analytic functions. Duke Math. J. **47**, 789–801.

[1984] Optimal sampling of holomorphic functions. Amer. J. Math. **106**, 593–609.

[1985] Optimal sampling of holomorphic functions. II. Math. Ann. **273**, 131–147.

Fisher, S. D., Stessin, M. I.

[1991] The n-width of the unit ball of H^q. J. Approx. Theory **67**, 347–356.

[1994a] The n-width of the unit ball of H^q — addendum. J. Approx. Theory **79**, 167–168.

[1994b] n-widths of classes of holomorphic functions with reproducing kernels. Illinois J. Math. **38**, 589–615.

Forst, W.

[1977] Über die Brite von Klassen holomorpher periodisher Funktionen. J. Approx. Theory **19**, 325–331.

Galeev, E. M.

[1990] Bernstein diameters of infinite dimensional ellipsoids. In: Complex Analysis. Differential and Integral Equations, Kalmyk State University, Elista, pp.14–20 (in Russian).

Ganelius, T. H.

[1976] Rational approximation in the complex plane and on the line. Ann. Acad. Sci. Fenn. Ser. A. I. **2**, 129–145.

Garnett, J. B.

[1981] Bounded Analytic Functions. Academic Press, New York.

Gohberg, I. C., Krein, M. G.

[1965] Introduction to the Theory of Linear Nonselfadjoint Operators. Nauka, Moscow (Russian). English edition: Transl. Math. Monographs, vol. 18. Amer. Math. Soc., Providence, RI, 1969.

Goluzin, G. M.

[1966] Geometric Theory of Functions of a Complex Variable. Nauka, Moscow (Russian). English edition: Amer. Math. Soc., Providence, RI, 1969.

Gonchar, A. A.

[1969] On Zolotarev problems connected with rational functions. Mat. Sb. **78**, 640–654. English transl. in Math. USSR Sb. **7** (1969), 623–635.

Hutton, C. V., Morrell, J. S., Retherford, J. R.

[1976] Diagonal operators, approximation numbers, and Kolmogoroff diameters. J. Approx. Theory **16**, 48–80.

Ioffe, A. D., Tikhomirov, V. M.

[1968] Duality of convex functions and extremal problems. Uspekhi Mat. Nauk **23**, 51–116. English transl. in Russian Math. Surveys **23** (1968), 53–124.

[1974] Theory of Extremal Problems. Nauka, Moscow (Russian). English edition: North–Holland, Amsterdam, 1979.

Ismagilov, R. S.

[1968] On n-dimensional diameters of compacts in a Hilbert space. Funktsional. Anal. i Prilozhen. **2**, 32–39. English transl. in Functional. Anal. Appl. **2** (1968), 125–132.

Khavinson, S. Ya.

[1963] The theory of extremal problems for bounded analytic functions satisfying additional conditions inside the domain. Uspekhi Mat. Nauk **18**, 25–98 (in Russian).

[1981] Foundation of the Theory of Extremal Problems with Additional Conditions. Moscow Civil Engineering Institute, Moscow. (Russian). English edition: Two Papers on Extremal Problems in Complex Analysis. Amer. Math. Soc. Translation. Ser. 2, vol. 129, Providence, RI, 1986.

Kolmogorov, A. N.

[1939] On inequalities between the upper bounds of the successive derivatives of an arbitrary function on an infinite interval. Uchenye Zap. MGU, Mat. **30** (3), 3–16. Amer. Math. Soc. Translation. Ser. 2, vol. 2, Providence, RI, 1962, 233–243.

Kolmogorov, A. N., Petrov, A. A., Smirnov, Yu. M.

[1947] A formula of Gauss in the theory of the method of least squares. Izv. Akad. Nauk SSSR. Ser. Mat. **11**, 561–566 (in Russian).

Korneichuk, N. P.

[1976] Extremal Problems of Approximation Theory. Nauka, Moscow (Russian).

[1987] Exact Constants in Approximation Theory. Nauka, Moscow (Russian). English edition: Cambridge University Press, 1991.

Loeb, H. L.

[1974] A note on optimal integration in H_∞. C. R. Acad. Bulgare Sci. **27**, 615–619.

Loeb, H. L., Werner, H.

[1974] Optimal numerical quadrature in H_p spaces. Math. Z. **138**, 111–117.

Macintyre, A. J., Rogosinski, W. W.

[1950] Extremum problems in the theory of analytic functions. Acta Math. **82**, 275–325.

Magaril-Il'yaev, G. G., Osipenko, K. Yu.

[1991] Optimal recovery of functionals based on inaccurate data. Mat. Zametki **50**, 85–93. English transl. in Math. Notes **50** (1991), 1274–1279.

Magaril-Il'yaev, G. G., Tikhomirov, V. M.

[1997] Kolmogorov-type inequalities for derivatives. Mat. Sb. **188** (1997), 73–106. English transl. in Sbornic: Mathematics **188** (1997), 1799–1832.

Mal'cev, A. I.

[1947] A remark on the paper of A. N. Kolmogorov, A. A. Petrov, and Yu. M. Smirnov "A formula of Gauss on the theory of least squares". Izv. Akad. Nauk SSSR. Ser. Mat. **11**, 567–568 (in Russian).

Marchuk, A. G., Osipenko, K. Yu.

[1975] Best approximation of functions specified with an error at finite number of points. Mat. Zametki **17**, 359–368. English transl. in Math. Notes **17** (1976), 207–212.

Melkman, A. A., Micchelli, C. A.

[1979] Optimal estimation of linear operators in Hilbert spaces from inaccurate data. SIAM J. Numer. Anal. **16**, 87–105.

Micchelli, C. A., Rivlin, T. J.

[1977] A survey of optimal recovery. In: Optimal Estimation in Approximation Theory, C. A. Micchelli, T. J. Rivlin (eds.), Plenum Press, New York, pp.1–54.

[1985] Lectures on Optimal Recovery. Lecture Notes in Mathematics, vol. 1129. Springer–Verlag, Berlin, pp.21–93.

Newman, D. J.

[1979] Quadrature formulae for H^p functions. Math. Z. **166**, 111–115.

Nikolskii, S. M.

[1946] Approximation of functions in the mean by trigonometrical polynomials. Izv. Akad. Nauk SSSR, Ser. Mat. **10**, 207–256 (in Russian).

Osipenko, K. Yu.

[1972] Optimal interpolation of analytic functions. Mat. Zametki **12**, 465–476. English transl. in Math. Notes **12** (1972), 712–719.

[1976] Best approximation of analytic functions from their values at a finite number of points. Mat. Zametki **19**, 29–40. English transl. in Math. Notes **19** (1976), 17–23.

[1982] Best methods for approximating analytic functions given with an error. Mat. Sb. **118**, 350–370. English transl. in Math. USSR Sb. **46** (1983), 353–374.

[1985] The Heins problem and optimal extrapolation of analytic functions given with error. Mat. Sb. **126**, 566–575. English transl. in Math. USSR Sb. **54** (1986), 566–575.

[1988] On best and optimal quadrature formulas on classes of bounded analytic functions. Izv. Akad. Nauk SSSR. Ser. Mat. **52**, 79–99. English transl. in Math. USSR Izv. **32** (1989), 77–97.

[1990] Blaschke products which have the least deviation. Mat. Zametki **47**, 71–80. English transl. in Math. Notes **47** (1990), 471–477.

[1991] Best and optimal recovery methods for classes of harmonic functions. Mat. Sb. **182**, 723–745. English transl. in Math. USSR Sb. **73** (1992), 111–133.

[1994a] The Carathéodory–Fejér problem and optimal recovery of derivatives in Hardy spaces. Mat. Sb. **185**, 27–42. English transl. in Russian Acad. Sci. Sb. Math. **81** (1995), 21–33.

[1994b] Inequalities for derivatives of functions analytic in a strip. Mat. Zametki **56**, 114–122. English transl. in Math. Notes **56** (1994), 1069–1074.

[1994c] On n-widths, optimal quadrature formulas, and optimal recovery of functions analytic in a strip. Izv. Ross. Akad. Nauk. Ser. Mat. **58**, 55–79. English transl. in Russian Acad. Sci. Izv. Math. **45** (1995), 55–78.

[1995a] Exact values of n-widths and optimal quadratures on classes of bounded analytic and harmonic functions. J. Approx. Theory **82**, 156–175.

[1995b] On N-widths of holomorphic functions of several variables. J. Approx. Theory **82**, 135–155.

[1996] Optimal recovery of periodic functions from Fourier coefficients given with an error. J. Complexity **12**, 35–46.

[1997a] Exact n-widths of Hardy–Sobolev classes. Constr. Approx. **13**, 17–27.

[1997b] On the precise values of n-widths for classes defined by cyclic variation diminishing operators. Mat. Sb. **188**, 113–126. English transl. in Sbornic: Mathematics **188**, 1371–1383.

[1999] Optimal recovery of the derivative of periodic analytic functions from Hardy classes. J. Approx. Theory **97**, 384–395.

Osipenko, K. Yu., Parfenov, O. G.

[1995] Ismagilov type theorems for linear, Gel'fand and Bernstein n-widths. J. Complexity **11**, 474–492.

Osipenko, K. Yu., Stessin, M. I.

[1990] On n-widths of the Hardy class H^2 in the unit ball of C^n. Uspekhi Mat. Nauk **45**, 193–194. English transl. in Russian Math. Surveys **45** (1990), 235-236.

[1991] On problems of recovery in Hardy and Bergman spaces. Mat. Zametki **49**, 95–104; English transl. in Math. Notes **49** (1991), 395–401.

[1992a] On optimal recovery of a holomorphic function in the unit ball of C^n. Constr. Approx. **8**, 141–159.

[1992b] On some problems of optimal recovery of analytic and harmonic functions from inaccurate data. J. Approx. Theory **70**, 206–228.

[1993] Optimal recovery of the derivatives of bounded analytic and harmonic functions from inaccurate data. Mat. Zametki **53**, 87–97. English transl. in Math. Notes **53** (1993), 513–520.

Osipenko, K. Yu., Wilderotter, K.

[1997] Optimal information for approximating periodic functions. Math. Comput. **66**, 1579–1592.

Ovchintsev, M. P.

[1989a] A best method of approximation of regular bounded functions in an annulus from their values at given points. Izv. Vysh. Uchebn. Zaved. Mat. **5**, 32–39. English transl. in Soviet Math. (Iz. VUZ) **33** (1989).

[1989b] On the optimal recovery of functions of the class E_p in an annulus. Sibirsk. Mat. Zh. **30**, 87–101. English transl. in Siberian Math.

J. **30** (1989), 568–580.

Parfenov, O. G.

[1985] Gel'fand diameters of the unit ball of the Hardy class H^p in weight spaces. Mat. Zametki **37**, 171–175. English transl. in Math. Notes **37** (1985), 96–98.

[1994a] Bernstein widths of some classes of analytic functions. Mat. Zametki **56**, 123–131. English transl. in Math. Notes **56** (1994), 1075–1081.

[1994b] Gel'fand widths of certain classes of analytic functions. Mat. Sb. **185**, 105–112. English transl. in Russian Acad. Sci. Sb. Math. **82** (1995), 223–229.

Pinkus, A.

[1979] On n-widths of periodic functions. J. Analyse Math. **35**, 209–235.

[1985] n-Widths in Approximation Theory. Springer–Verlag, Berlin.

Plaskota, L.

[1996] Noisy Information and Computational Complexity. Cambridge University Press.

Rivlin, T. J., Ruscheweyh, St., Shaffer, D., Wirths. K. J.

[1983] Optimal recovery of the derivative of bounded analytic functions. IMA J. of Numer. Anal. **3**, 327–332.

Rivlin, T. J., Shaffer, D.

[1983] Optimal estimation of the derivative of bounded analytic functions. IBM Reserch Report, RC 9843.

Rogosinski, W. W., Shapiro, H. S.

[1953] On certain extremum problems for analytic functions. Acta Math. **90**, 287–318.

Rudin, W.

[1980] Function Theory in the Unit Ball of C^n. Springer–Verlag, New York.

Smolyak, S. A.

[1965] On Optimal Restoration of Functions and Functionals of Them. Candidate dissertation. Moscow State University, Moscow (Russian).

Sofman, L. B.

[1969] Diameters of octahedra. Mat. Zametki **5**, 429–436. English transl. in Math. Notes **5** (1969), 258–262.

[1973] Diameters of an infinite-dimensional octahedra. Vestnik Mosk. Univ. Mat. **28**, 54–56. English transl. in Moscow Univ. Math. Bull. **28** (1973), 45–47.

Stechkin, S. B.

[1954] On the best approximation of given classes by any polynomials. Uspekhi Mat. Nauk **9**, 133-134 (in Russian).

Stenger, F.

[1978] Optimal convergence of minimum norm approximation in H_p. Numer. Math. **29**, 345–362.

Taikov, L. V.

[1967] On the best approximation in the mean of certain classes of analytic functions. Mat. Zametki **1**, 155–162. English transl. in Math. Notes **1** (1967), 104–109.

Tikhomirov, V. M.

[1960] Diameters of sets in function spaces and the theory of best approximations. Uspekhi Mat. Nauk **15**, 81–120. English transl. in Russian Math. Surveys **15** (1960), 75–111.

[1976] Some Questions of Approximation Theory. Moscow State University, Moscow (Russian).

[1987] Approximation Theory. In: Itogi Nauki i Tekhniki. Sovremennye Problemy Mat. Fundamental'nye Napravlenia, vol. 14, VINITI, Moscow, pp.103–260. English edition: Encyclopedia of Math. Sci., vol. 14, Springer–Verlag, Berlin, 1990.

Tikhomirov, V. M., Magaril-Il'yaev, G. G.

[1985] Inequalities for derivatives. In: Selected Works of A. N. Kolmogorov. Mathematics and Mechanics, Nauka, Moscow, pp.252–261. English edition: Kluwer, Dordrecht, 1991.

Traub, J. F., Woźniakowski, H.

[1980] A General Theory of Optimal Algorithms. Academic Press, New York.

Traub, J. F., Wasilkowski, G. W., Woźniakowski, H.

[1988] Information-Based Complexity. Academic Press, New York.

Walsh, J. L.

[1965] Interpolation and Approximation by Rational Functions in the Complex Plane. Amer. Math. Soc. Colloq. Publ. XX. Amer. Math. Soc., Providence, RI.

Widom, H.

[1972] Rational approximation and n-dimensional diameters. J. Approx. Theory **5**, 343–361.

Wilderotter, K.

[1992a] n-widths of H^p-spaces in $L_q(-1,1)$. J. Complexity **8**, 324–335.

[1992b] Optimal recovery of bounded analytic functions in an annulus. In: Optimal Recovery, B. Bojanov, H. Woźniakowski (eds.), Nova Science Publishers, New York, pp.311–323.

[1995] The fine structure of the n-widths of H^p-spaces in $L_q(-1,1)$. J. Complexity **11**, 456–460.

[1996] Optimal approximation of periodic analytic functions with integrable boundary values. J. Approx. Theory **84**, 236–246.

Author Index

Subject Index